STRATEGIES FOR A BEA SATELLITE HEALTH CARE ACCOUNT

SUMMARY OF A WORKSHOP

Christopher Mackie, *Rapporteur*

Committee on National Statistics

Division of Behavioral and Social Sciences and Education

NATIONAL RESEARCH COUNCIL
OF THE NATIONAL ACADEMIES

THE NATIONAL ACADEMIES PRESS
Washington, D.C.
www.nap.edu

THE NATIONAL ACADEMIES PRESS 500 Fifth Street, N.W. Washington, DC 20001

NOTICE: The project that is the subject of this report was approved by the Governing Board of the National Research Council, whose members are drawn from the councils of the National Academy of Sciences, the National Academy of Engineering, and the Institute of Medicine. The members of the committee responsible for the report were chosen for their special competences and with regard for appropriate balance.

This study was supported by Contract No. DG132106CN028 between the National Academy of Sciences and the United States Department of Commerce, Bureau of Economic Analysis. The work of the Committee on National Statistics is provided by a consortium of federal agencies through a grant from the National Science Foundation (Number SBR-0112521). Any opinions, findings, conclusions, or recommendations expressed in this publication are those of the author(s) and do not necessarily reflect the views of the organizations or agencies that provided support for the project.

International Standard Book Number-13: 978-0-309-12717-2
International Standard Book Number-10: 0-309-12717-3

Additional copies of this report are available from The National Academies Press, 500 Fifth Street, NW, Lockbox 285, Washington, DC 20055; (800) 624-6242 or (202) 334-3313 (in the Washington metropolitan area); Internet, http://www.nap.edu.

Copyright 2009 by the National Academy of Sciences. All rights reserved.

Printed in the United States of America

Suggested citation: National Research Council. (2009). *Strategies for a BEA Satellite Health Care Account: Summary of a Workshop*. Christopher Mackie, Rapporteur. Steering Committee for the Workshop to Provide Guidance for Development of a Satellite Health Care Account at the Bureau of Economic Analysis. Committee on National Statistics, Division of Behavioral and Social Sciences and Education. Washington, DC: The National Academies Press.

THE NATIONAL ACADEMIES
Advisers to the Nation on Science, Engineering, and Medicine

The **National Academy of Sciences** is a private, nonprofit, self-perpetuating society of distinguished scholars engaged in scientific and engineering research, dedicated to the furtherance of science and technology and to their use for the general welfare. Upon the authority of the charter granted to it by the Congress in 1863, the Academy has a mandate that requires it to advise the federal government on scientific and technical matters. Dr. Ralph J. Cicerone is president of the National Academy of Sciences.

The **National Academy of Engineering** was established in 1964, under the charter of the National Academy of Sciences, as a parallel organization of outstanding engineers. It is autonomous in its administration and in the selection of its members, sharing with the National Academy of Sciences the responsibility for advising the federal government. The National Academy of Engineering also sponsors engineering programs aimed at meeting national needs, encourages education and research, and recognizes the superior achievements of engineers. Dr. Charles M. Vest is president of the National Academy of Engineering.

The **Institute of Medicine** was established in 1970 by the National Academy of Sciences to secure the services of eminent members of appropriate professions in the examination of policy matters pertaining to the health of the public. The Institute acts under the responsibility given to the National Academy of Sciences by its congressional charter to be an adviser to the federal government and, upon its own initiative, to identify issues of medical care, research, and education. Dr. Harvey V. Fineberg is president of the Institute of Medicine.

The **National Research Council** was organized by the National Academy of Sciences in 1916 to associate the broad community of science and technology with the Academy's purposes of furthering knowledge and advising the federal government. Functioning in accordance with general policies determined by the Academy, the Council has become the principal operating agency of both the National Academy of Sciences and the National Academy of Engineering in providing services to the government, the public, and the scientific and engineering communities. The Council is administered jointly by both Academies and the Institute of Medicine. Dr. Ralph J. Cicerone and Dr. Charles M. Vest are chair and vice chair, respectively, of the National Research Council.

www.national-academies.org

STEERING COMMITTEE FOR THE WORKSHOP TO PROVIDE GUIDANCE FOR DEVELOPMENT OF A SATELLITE HEALTH CARE ACCOUNT AT THE BUREAU OF ECONOMIC ANALYSIS

JOSEPH P. NEWHOUSE (*Chair*), Division of Health Policy Research and Education, Harvard University
BARBARA FRAUMENI, Muskie School of Public Service, University of Southern Maine
GAIL WILENSKY, Project HOPE, Bethesda, Maryland

CHRISTOPHER MACKIE, *Study Director*
MICHAEL SIRI, *Senior Program Assistant*

COMMITTEE ON NATIONAL STATISTICS
2007-2008

WILLIAM F. EDDY (*Chair*), Department of Statistics, Carnegie Mellon University
KATHARINE ABRAHAM, Department of Economics and Joint Program in Survey Methodology, University of Maryland
WILLIAM DUMOUCHEL, Lincoln Technologies, Inc., Waltham, Massachusetts
JOHN HALTIWANGER, Department of Economics, University of Maryland
V. JOSEPH HOTZ, Department of Economics, Duke University
KAREN KAFADAR, Department of Statistics, Indiana University
DOUGLAS MASSEY, Department of Sociology, Princeton University
SALLY MORTON, RTI International, Research Triangle Park, North Carolina
VIJAY NAIR, Department of Statistics and Department of Industrial and Operations Engineering, University of Michigan
JOSEPH P. NEWHOUSE, Division of Health Policy Research and Education, Harvard University
SAMUEL H. PRESTON, Population Studies Center, University of Pennsylvania
KENNETH PREWITT, School of International and Public Affairs, Columbia University
LOUISE RYAN, Department of Biostatistics, Harvard University
ROGER TOURANGEAU, Joint Program in Survey Methodology, University of Maryland, and Survey Research Center, University of Michigan
ALAN ZASLAVSKY, Department of Health Care Policy, Harvard University Medical School

CONSTANCE F. CITRO, *Director*

Preface

The workshop summarized in this report was convened by the Committee on National Statistics (CNSTAT) on behalf of the Bureau of Economic Analysis (BEA) to advance their plans to produce a satellite health care account designed to improve measurement of activity in the health and medical care sectors of the economy. The purpose of this workshop was to elicit expert guidance on strategies for better coordinating health expenditure statistics, developing medical care price measures, and producing comprehensive sets of accounts for health care sector income, expenditure, and product organized by type of disease that are more directly useful for measuring health care inputs, outputs, and productivity. Such data are central to tracking the share of the country's resources devoted to medical care, how rapidly that share is increasing, and, most importantly, what the country is getting from this increased spending on medical care.

The workshop consisted of sessions covering the following topics: (1) plans for a satellite health care account, for which Ana Aizcorbe of BEA presented the goals of the agency's health accounting program and progress to date on the project; (2) the Altarum Institute's construction of nominal expenditures by disease, for which Charles Roehrig presented research findings; (3) price indexes and volume measures, for which Ralph Bradley and Bonnie Murphy of the Bureau of Labor Statistics discussed new work on health care indexes in the Consumer Price Index and Producer Price Index programs; (4) measuring treatment outcomes, for which Mark McClellan of the Brookings Institution discussed the challenges of constructing measures of treatment outcomes, essential for monitoring the quality of medical care; and (5) national accounting issues, for which Brian Moyer of BEA provided details about BEA's plan to modify the industry side of the national accounts so that they can remain synchronized with the disease-based organizational structure proposed for the satellite program.

On behalf of CNSTAT, I thank all of the workshop presenters for preparing detailed presentations and background papers, which provided the basis for engaging and productive discussion. I also thank the workshop participants for their insightful comments and fruitful exchange of ideas, as well as for their input as staff drafted this report. I especially wish to thank my colleagues on the workshop steering committee, Barbara Fraumeni and Gail Wilensky, for their helpful guidance and leadership in planning and moderating the workshop.

We are grateful for BEA's sponsorship of the workshop and thank Dennis Fixler, Steven Landefeld, and, in particular, Ana Aizcorbe for offering guidance to staff and the workshop steering committee in the development of the agenda and in identifying the workshop goals. David Cutler, Dennis Fryback, Alan Garber, Emmett Keeler, Allison Rosen, and Jack Triplett, serving on CNSTAT's Committee on National Health Accounts, which is working on another closely related project funded by the National Institute on Aging, provided input to the workshop planning as well.

Michael Siri of the CNSTAT staff expertly managed the administrative details and workshop arrangements and worked on the report itself. Connie Citro, director of CNSTAT, provided guidance and support throughout the project. Kirsten Sampson Snyder guided this report through the review process. Most importantly, Christopher Mackie, the staff study director for the workshop, took the lead in planning the workshop and prepared and revised the report on the basis of comments from reviewers and workshop participants.

This report has been reviewed in draft form by individuals chosen for their diverse perspectives and technical expertise, in accordance with procedures approved by the Report Review Committee of the National Research Council. The purpose of this independent review is to provide candid and critical comments that assist the institution in making its report as sound as possible, and to ensure that the report meets institutional standards for objectivity, evidence, and responsiveness to the study charge. The review comments and draft manuscript remain confidential to protect the integrity of the deliberative process.

The panel thanks the following individuals for their review of this report: Jessica Banthin, Center for Financing, Access, and Cost Trends, Agency for Healthcare Research and Quality; Barbara M. Fraumeni, Muskie School of Public Service, University of Southern Maine, Portland; Dale W. Jorgenson, Department of Economics, Harvard University; and Matthew D. Shapiro, Department of Economics, University of Michigan. Although the reviewers have provided many constructive comments and suggestions, they were not asked to endorse the conclusions or recommendations, nor did they see the final draft of the report before its release. The review of this report was overseen by Katharine G. Abraham, Joint Program in Survey Methodology, University of Maryland. Appointed by the National Research Council, she was responsible for making certain that the independent examination of this report was carried out in accordance with institutional procedures and that all review comments (including her own) were carefully considered. Responsibility for the final content of the report rests entirely with the institution.

> Joseph P. Newhouse, *Chair*
> Steering Committee for the Workshop to Provide
> Guidance for Development of a Satellite Health
> Care Account at the Bureau of Economic Analysis

Contents

1 INTRODUCTION   1
    1.1. Project Description and Report Structure, 1
    1.2. What Kind of "Satellite" Health Care Account?, 4
    1.3. Objectives of the BEA Project—A Staged Strategy for Developing a National Health Care Account, 7

2 ALLOCATING NOMINAL EXPENDITURES ON MEDICAL CARE: A DISEASE-BASED CONCEPTUAL APPROACH   12
    2.1. Methods for Attributing Spending to Treatments: The Big Comorbidity Issue, 13
    2.2. Allocating Personal Health Expenditures by Medical Condition: The Altarum Institute Project, 18
    2.3. Comparing the Methods, 22
    2.4. Data Needs for Expenditure Accounting, 25

3 PRICE INDEXES: CALCULATING REAL MEDICAL CARE GDP   30
    3.1. Pricing Treatments to Capture Changing Technologies, Input Substitution, and Population Heterogeneity, 32
    3.2. BEA's Strategy for Coordinating the Industry Input Accounts with the Disease Treatment–Based Expenditure Concept, 35
    3.3. Tracking Quality Change of Medical Goods and Services, 41
    3.4. The Role of the BLS Price Indexes, 44
    3.5. Outcomes and Quality Change, 56
    3.6. Data Needs for Price Measurement, Tracking Outcomes, and Quality Adjustment, 60

4 SUMMARY, PERSPECTIVE, AND PROSPECTS FOR MOVING FORWARD 63

REFERENCES 66

APPENDIXES

A Summary Statistics from the Medical CPI and U.S. Medical Expenditures Panel Survey 67
B Workshop Agenda and Participants 84
C Adapting BEA's National and Industry Accounts for a Health Care Satellite Account 88
    Brent R. Moulton, Brian C. Moyer, and Ana Aizcorbe

1

Introduction

1.1. PROJECT DESCRIPTION AND REPORT STRUCTURE

In March 2008 the Committee on National Statistics of the National Academies held a workshop to assist the Bureau of Economic Analysis (BEA) with next steps as it develops plans to produce a satellite health care account designed to improve its measurement of economic activity in the medical care sector and to be useful for health care policy. The ultimate objectives of the BEA program are to compile medical care spending information by type of disease that is more directly useful for measuring health care inputs, outputs, and productivity than are current estimates of spending by type of provider; produce a comprehensive set of accounts for health care–sector income, expenditure, and product; develop medical care price and real output measures that better break out changes in the delivery of health care from changes in the prices of that care; and coordinate BEA and Centers for Medicare and Medicaid Services (CMS) health expenditure statistics.

BEA, at this point, still very much in the research stage of this project, has committed to produce a detailed proposal for a satellite or experimental account by the end of 2009. The agency is coordinating its work with other efforts, notably a project led by David Cutler (Harvard University) and Allison Rosen (University of Michigan), and through collaboration with colleagues at other statistical agencies that has taken place during the past 1-2 years. The hope at BEA is that work associated with this project will eventually lead to methodological advances in the national income and product accounts (NIPAs) by improving the conceptual basis for accurately measuring prices and quantities of the economy's medical care goods and services.

The steering committee for the workshop consisted of Barbara Fraumeni (University of Southern Maine, and formerly BEA), Joseph Newhouse (Harvard University), and Gail Wilensky (Project HOPE). The BEA-sponsored workshop included sessions covering the following topics:

- **Plans for a Satellite Health Care Account.** Ana Aizcorbe (chief economist, BEA) presented the goals of BEA's health accounting program, progress to date on the project, and a summary of BEA's strategies for dealing with key measurement issues and data needs. Dale Jorgenson (Harvard University) and Matthew Shapiro (University of Michigan) served as discussants.
- **Constructing Nominal Expenditures by Disease.** Charles Roehrig (Altarum Institute) provided an overview of work by him and his colleagues developing time-series estimates of national expenditures by medical condition. This type of data will be essential as health accounting projects move forward because the treatment for a specific condition or disease provides an organizing principle for defining units of service in the medical care sector. David Cutler and Allison Rosen were discussants.
- **Price Indexes and Volume Measures.** Ralph Bradley and Bonnie Murphy (Bureau of Labor Statistics, Consumer Price Index and Producer Price Index programs, respectively) discussed their agency's plans to research and generate price indexes organized by broad disease category. Price indexes, which are used to decompose changes in current dollar estimates into price and quantity components, are essential for calculating real gross domestic product (GDP) for the various sectors of the economy. The Producer Price Index (PPI) program has developed a method to quality adjust its current hospital indexes by using quality indicators contained in the CMS Hospital Compare database. The Consumer Price Index (CPI) program is generating experimental price indexes, also organized by major disease category, by merging medical expenditure and utilization data from the Medical Expenditure Panel Survey with the CPI production database of the Bureau of Labor Statistics (BLS). Jack Triplett (Brookings Institution) and Patricia Danzon (University of Pennsylvania) were discussants.
- **Measuring Treatment Outcomes.** Mark McClellan (Brookings Institution) discussed the challenges of constructing measures of treatment outcomes, which are essential for monitoring the quality of medical care and, in turn, changes in the real output of the sector. He also provided an assessment of the current state of knowledge evident in the outcomes research literature and of how that information is influencing relevant policy discussions.

- **National Accounting Issues.** BEA discussed national accounting issues that must be resolved in order to produce a satellite account for health. These include how to construct measures of real expenditures for health care industries, define disease and product classes, and develop a set of deflators for medical care industries; the importance of sources of payment data was also discussed. The central presentation of the session, by Brian Moyer of BEA, focused on development of an approach for coordinating the spending and industry sides of the national accounts. Barbara Fraumeni and Sherry Glied (Columbia University) were the discussants.
- **Summary comments and directions** were provided by Gail Wilensky.

Throughout the meeting, participants demonstrated enthusiastic support for the BEA effort and offered encouragement to the project leaders. Dale Jorgenson referred to BEA's work on the project as landmark, noting that it could have a significant impact—and not just on measured GDP, but also on the way people think about health and the health care sector. He stated that it is promising that BEA had concentrated talent and resources on this important research topic.

Matthew Shapiro agreed, observing that the statistical system appears poised to make major progress on health accounting issues, and added that the workshop occurred at a pivotal time. Both the BLS and BEA are currently engaged in innovative research that will enhance their ability to provide comprehensive data for measuring activity in the medical care sector of the economy. Shapiro expressed hope that participants in the workshop, and others, would be able to leverage two elements: The first is to build on the conceptual work by pioneering researchers—such as Scitovsky (1967); Cutler, McClellan, Newhouse, and Remler (1998); and Berndt, Busch, and Frank (1998)—that sought to measure prices and quantities of medical care using a disease treatment framework; the second is harnessing the potential of the tremendous amount of data available to implement this kind of measurement on a production-level basis. Academic researchers began the work, disease case by disease case, but the systematic production of data by a statistical agency would mark a huge step forward.

Finally, CNSTAT director Constance Citro offered her encouragement and expressed hope that the workshop would help BEA push forward in the development of a satellite health care account. She also noted the complementary work by a CNSTAT panel seeking to advance foundations for a data system that would relate both medical care and other inputs to incremental changes in the health of the population. That project is funded by the National Institute on Aging and, like this workshop, is also chaired by Joseph Newhouse. The two projects are related in that development of data on expenditures, prices, and quantities for medical care—BEA's program—is a key component of the broader health account concept as well.

1.2. WHAT KIND OF "SATELLITE" HEALTH CARE ACCOUNT?

The term "satellite account" has been used to describe different kinds of activities undertaken at statistical agencies, both in the United States and abroad. As Barbara Fraumeni put it, satellite accounts can show more detail than that present in the national accounts, or they can extend to areas not covered in the NIPAs at all. Research to improve the methodologies and data that underlie the NIPAs and GDP measurement is constantly active at BEA (and other statistical agencies as well); historically, satellite projects that entail experimental work intended to enhance the detail and accuracy of the accounts have been part of the effort. BEA's experimental research and development (R&D) satellite account is an example—it provides a more detailed look at the composition of R&D costs and a more accurate measure of capital investment.[1] In other instances, satellite projects have been designed to improve data on economic activities typically considered peripheral or even out of scope of the NIPAs and GDP. Environmental accounting is an example of this kind of work.[2]

Ana Aizcorbe kicked off the workshop by outlining the agency's plans for a satellite health care account. She began by defining the purpose of the project. Compared with work by academic health economists on data systems designed to track changes in population health alongside the investment of inputs to health, she described the BEA initiative as modest. BEA is most concerned with one input to health—medical care—because it is the piece that is most relevant to the market-oriented national accounts; it is also numerically important (medical care accounts for a large and growing portion of GDP), a key policy topic, and the component of health on which BEA is most likely to be able to make progress.

David Cutler supported BEA's approach, noting that, in phasing the project, there are tasks that BEA will clearly be able to do itself and others that will require collaboration. For example, BEA is not going to hire a staff of physicians to establish what is happening with patient outcome trends in order to assess the changing quality of medical services. That will involve academic work and, perhaps someday, other government departments, such as the Department of Health and Human Services. At several points during the proceedings, participants noted the importance of differentiating aspects of the research agenda that should take place inside BEA with those that should progress elsewhere.

---

[1] The satellite account also introduces a different conceptual approach to R&D spending, treating it as investment.

[2] The BEA website provides a wealth of information on the methodologies, content, and scope of the NIPAs; "Measuring the Economy: A Primer on GDP and the National Income and Product Accounts" (http://bea.gov/national/pdf/nipa_primer.pdf) provides a good starting point. Information is also available there about the various satellite accounts that have been developed over the years by the agency. Additional discussion of the scope and coverage of the NIPAs and the role of satellite accounts can be found in a CNSTAT report (National Research Council, 2005, pp. 14-19).

INTRODUCTION 5

A Broad Population Health Account

The scope of the national income and product accounts has traditionally been limited, with some exceptions, to activities that take place within the market. But, because it has long been recognized that a quantitatively significant amount of economic activity occurs beyond the market, efforts have been made to expand accounting structures to allow broader measurement of the economy. The history of economic accounting includes forays by statistical agencies into such projects with broader scope. One example is the satellite accounts that have been developed in various countries, including the United States, to estimate the economic contributions to society of the environment (or parts thereof). Another example is a household production account, with which several statistical agencies around the world have begun experimenting. The range of potential "GDP-expanding" areas of nonmarket accounting is catalogued in *Beyond the Market: Designing Nonmarket Accounts for the United States* (National Research Council, 2005).

For the topic at hand, the broad-scope satellite account would define population health as the output and would encompass nonmarket components beyond those that are included in the national economic accounts. Sherry Glied described the idea clearly: Medical care and an array of other variables would be organized into a kind of health production function, thereby providing a framework for measuring changes in health and the factors affecting that change. This output concept—designed to track the value of incremental changes in health, as well as inputs in the production of health, such as medical care, the environment, lifestyle choices, and other factors—is clearly broader in scope than that which now guides measurement of the medical care industry of the economy in the national accounts. Natural resource and environmental accounting have the same basic characteristic; there are additions (or subtractions) to GDP driven by inclusion of nonmarket elements.

During open discussion, Jorgenson identified the linkage between the medical care component of the NIPAs and this broader kind of population health account: Output of the medical sector is very important in GDP measurement but, for many aspects of health policy, medical care outcomes are of the greatest concern. The inputs of the broader account would be precisely the treatments that are the outputs of the market account. In other words, the medical sector produces treatments that are, in turn, inputs into the production of incremental improvements in health. So there is a conceptual framework that links industries and their inputs to treatments, and treatments to outcomes; in both cases, Jorgenson noted, it is useful to focus on treatments, because that is what would enter into the GDP. He went on to describe measurement of medical care outcomes as the $64 trillion question. Here, one must deal with the issue of how medical care produces improved health, if it does, which is very difficult.

The objective of the broader kind of account is, like standard market-oriented economic approaches, to accurately estimate prices and quantities. The quantity estimate is linked directly to changes in health outcomes. When there is improvement, people are able to enjoy longer, healthier lives, so there is an incremental change in individuals' lifetime "income," broadly defined, as a result of medical treatment. Finding the part of that improvement in health that is attributable to medical care is a very important research task. Jorgenson noted that workshop attendees had made numerous important contributions in this area; the most prominent being the disease-specific studies, such as the one on heart attacks by Cutler, McClellan, Newhouse, and Remler (1998).

Jorgenson concluded that the nonmarket component of health accounting is presently at the research frontier. It is not something that BEA could, given the current state of the art, fruitfully think of as part of its satellite system now under development; but, he advised, these ideas and goals—particularly as they relate to measurement of the changing quality of medical care—should be kept in mind at all times as part of the long-run objective.

The BEA Medical Care Account: Experimental Work to Improve the Market-Oriented NIPAs

Jorgenson suggested that perhaps the satellite account concept was not quite the way to think about BEA's project. He considers the broader health accounts (described above) a "satellite" to the NIPAs mainly because, in such a system, a nonmarket component of output is present—the incremental value of better health—that supplements the treatment of output in the national accounts. Fraumeni expressed the view that, independent of whether or not it is considered satellite work, the key substantive point is that BEA's objective initially should be to improve measurement of the medical care component of the national accounts rather than to construct a new set of accounts defining a different boundary of goods and services. She characterized BEA's research as focusing on a different way of measuring the prices of medical care. This is critical because new price indexes mean different deflators, which, in turn, affect estimates of real medical care GDP. Aizcorbe agreed with this characterization—that the agency is trying to fix the price deflator, which requires supporting work that is experimental.

Because price measurement work is central to the BEA program, other statistical agencies are involved. Triplett noted that, for in-scope market items, any new pricing methods that are developed can be taken back to BLS as a better way of doing things for certain purposes. He cited price adjustment methods for quality change in the computer index as an example. During the 1980s, BEA calculated the computer index using new methods that were then brought over to BLS; sometimes the methodological improvements have flowed the other way.

As part of this new approach to the medical care component of the NIPAs, BEA would need to develop a technique so that the revised prices associated with expenditures are reflected in the industry accounts as well. As Triplett put it, the measurement issues can be viewed either from the demand side (the CPI, for example, is constructed from the viewpoint of the consumer) or from

INTRODUCTION

the production or supply side, at the industry or subsector level. In price index theory, those two sides sometimes yield different measurement implications, and he asserted that this will also be the case for health care. Newhouse stated that what BEA's work is really doing is changing the good that is being priced. That affects the price index for consumer expenditures, which carries over to the product accounts. Both sides of the account—expenditures on outputs and costs of inputs—must, by definition, balance. Fraumeni made this point: An input-output framework underlies the NIPAs and, to make the new expenditure structure agree with the industry side, a new industry piece has to be created (how this piece is to be created is the topic of discussion in Section 3.2.).

The main point here is that BEA's upgraded account will not involve a revision of scope—it will be a system that stays within the market-oriented boundaries of the national accounts and that does not explicitly extend to a nonmarket component. Jorgenson agreed with this strategy, stating that BEA's new work will provide an alternative way of measuring medical care prices and it is (and should be) limited to that. BEA director Steven Landefeld provided additional clarity about the objectives of the program. He reiterated the point that, historically, satellite accounts have been pursued for different purposes—one being to expand the scope of the accounts, another to provide more incremental detail or to modify the structure within scope. It is the latter that the agency is pursuing, initially in an experimental context, with the new disease-based organizing approach. BEA's motivation is to improve measurement within the existing NIPAs, and Landefeld noted that this work could eventually affect overall measured rates of inflation and real GDP. Every time a better price index is introduced, it changes the deflator used to calculate real GDP, even if it does not change the categories or the scope of the accounts.

Landefeld cited BEA's travel and transportation satellite account as an example of work that takes place within the scope of the existing accounts but that breaks out additional detail for what is already covered. As in this work, which produces some fairly experimental numbers, he agreed that the agency needs to be more explicit in distinguishing between what will (or could) and will not affect official GDP estimates. He reiterated that BEA's work on medical care is, at this point, not geared toward expanding the scope of the accounts, and that the agency would leave it to others to conduct the cutting-edge research on that front and on the issues that directly affect the public health policy debate. Indeed, the reason that the agency has explicitly partnered with Cutler and Rosen is because it is not in a position to handle the expansion of scope side of the research agenda.

1.3. OBJECTIVES OF THE BEA PROJECT—A STAGED STRATEGY FOR DEVELOPING A NATIONAL HEALTH CARE ACCOUNT

The remaining chapters of this report summarize the day's proceedings. The views expressed by workshop participants are organized topically, more or less following the logical flow of process steps that may take place as BEA's work on health accounting unfolds. Essentially, BEA presented its plan, then participants provided critique, feedback, and detailed guidance on the plan. A sketch outline of BEA's project phases (and the rest of this report) follows.

Categorizing Nominal Expenditures on Medical Care: Using a Disease-Based Framework

The first major topic, and the one on which BEA is currently focusing most of its effort, is to rework the way expenditures on medical care goods and services are organized. The idea is to identify units of output that are meaningful from a consumer standpoint—and to get price and quantity measures for those units right. For example, a patient is not typically interested in buying an hour of a doctor's time or a day in a hospital; rather, a patient seeks treatment for a particular condition or ailment—the consumer wants to buy improved health. Workshop participants identified several potential units of measurement, such as a patient encounter or an episode, and considered the strengths and weaknesses of each. In Chapter 2, we describe characteristics of a system, as envisioned by workshop participants, for defining expenditure categories or "buckets" in which to allocate dollars spent in the economy on medical care.

One of the most difficult issue that arises is comorbidity; when patients require medical care for multiple conditions, it becomes much more complicated to assign expenditures to predefined categories accurately. For a seemingly simple event—a visit to the doctor—even at the claim level, there can be up to four diagnoses. So the question becomes how to make the judgment call in these cases: for the depressed patient with back pain who visits the doctor, should dollars spent be apportioned to the back pain or to the depression category? As described in this report, workshop participants considered and discussed several different methods—first listed diagnosis, grouper algorithms, person-based, etc.—for parsing spending in the presence of more than one condition.

During the opening presentation, Aizcorbe laid out the approaches that BEA is currently studying for allocating medical care spending by disease and for handling the comorbidity problem. She noted that it is not yet clear which one will be most appropriate for the national accounts. So far, much of BEA's work has relied on "grouper" programs that sort spending using claims data and thus is based on the patient's history. These programs allocate the dollars recorded on each claim into particular disease categories. Aizcorbe stated that, even if this does not turn out to be the option that is ultimately used, BEA would want to be able to answer the question, "Why weren't episode groupers used for parsing expenditures by disease?" The common complaint among workshop participants about grouper programs is that, at this point, they are constructed in a methodological black box; they are proprietary, and it is not clear if or when BEA would be given access to their internal workings. BEA participants expressed the view that, once the unit of output is defined indicating how expenditures are to be parsed, pricing medical care inputs (doctors' time, hospital services, drugs, medical equipment, etc.) will pose less severe conceptual problems. There are exceptions, however. For example, high-tech medical equipment is poorly measured in U.S. statistics, but this is not very different from other industry accounts in the NIPAs.

Another task on BEA's agenda is to continue work to reconcile the nominal spending estimates in the NIPAs with those reported by CMS in the national health expenditure accounts. In contrast to BEA's proposed framework, in which the focus is on organizing expenditure data by medical treatment categories for the full range of specifically defined diseases and conditions, CMS's accounts—which are the official estimates of health care spending for the United States—are designed to track final payments by type of provider. Although the data systems at the two agencies serve different purposes and therefore will not be identically structured, expenditure totals on medical care and for equivalently defined subcategories should align. Collaborative research is under way at CMS and BEA to further this objective.

Price Indexes: Calculating Real GDP for Medical Care

Many of the difficult tasks required for developing a satellite health care ccount relate to the calculation of price indexes needed to deflate nominal expenditure estimates for the measurement of real GDP. Aizcorbe expressed the view that the treatment-oriented price indexes that have been advocated by health economists are what should be used as deflators for the national accounts. For double-entry national accounts such as the NIPAs, deflators must be created for both the spending side (consisting of outputs purchased by consumers) and the industry side (consisting of inputs provided doctors, drug producers, hospitals, etc., in the production of medical care). Brian Moyer, whose presentation is summarized in Section 3.2., outlined BEA's strategy for modifying the structure of the industry accounts and for implementing the types of deflators that would be needed for the totals of each side to agree (or at least to minimize the statistical discrepancy between the two).

When a good is defined as the treatment of a disease or episode of illness, it necessarily means that a data series on spending by disease must be generated. Aizcorbe stated that figuring out how to use these data series to construct improved price deflators was a high priority for BEA. Sherry Glied, speaking from the perspective of a health economist (as opposed to an economic accountant), succinctly summarized the tasks at hand: Currently, many individual goods and services are delivered by different industries—doctors, hospitals, pharmaceutical companies, and so on. The plan is to continue to deflate those things using the PPI, or something similar, then to take all those individual components and create something called medical care, which is not any one treatment but a set of treatments for a disease. Doing that involves integrating services from different industries, pricing those bundles of services, then, ideally, capturing quality change associated with various medical innovations that may, among other things, allow substitution among medical care inputs. Items in this medical care box, then, will be deflated using price indexes organized along these disease treatment lines.

Index construction may draw from both the BLS's CPI and PPI programs, and BEA may introduce new price indexes specifically designed to meet the needs of the satellite account. The PPIs provided by BLS may turn out to be close to what is needed to deflate industry-side items, such as office visits, prescription drugs, and so on. However, the PPIs are not adequate for use on the spending side because the "good" demanded by consumers—treatments—does not align with this kind of delineation; rather, it combines these items.

Participants from BEA and BLS speculated that the differences between estimates based on the different kinds of indexes are numerically important. Preliminary research by Bradley and his colleagues (see Appendix A) found clear differences between indexes based on the PPI scope and the CPI scope. BEA has also done research and commissioned outside work using a different episode grouper that found significant variation as well.

Measuring Quality Change in Medical Care

Ideally, medical care price indexes used to deflate the economic accounts would be capable of reflecting changes in the quality of medical goods and services. In national income and product accounting, it is customary to adjust for instances when goods and services—in this case, treatments—are getting better or worse. Thinking seriously about how to measure changes in the quality and, in turn, the real cost of medical care requires monitoring information about outcomes associated with that care. Glied noted that, for this work, it seems logical to implement an episodes-of-treatment concept that reflects the way in which the medical profession—and not economists so much—measures outcomes. One way to compare quality across treaments, for example, is to examine clinical trials; while they have limitations, it would be counterproductive to construct episodes of treatment that did not match with outcome concepts that are being measuring in other fields.

BEA staff reported that they will postpone tackling the difficult quality change issue in the immediate future. The agency does not have medical expertise, so its strategy is to go after the things it believes can be fixed; later, research from the public health and health economics fields may advance enough to provide guidance about how to move forward on the quality measurement and outcomes pieces. Aizcorbe indicated that, even though BEA is not in a position to make major headway on the topic now—the program is still very much in the research stage—the agency realizes that accounting for quality change is an ultimate goal and that this need should be kept in mind as the satellite framework develops.

Other Issues: Source of Payment Data

Throughout the course of the day, a range of other topics was covered. One of these is data needs—especially those related to expenditure accounting and to price index development. Those working on these topics have, because of their complexity, discovered the need to draw from a broad array of data sources—aggregate and micro, longitudinal and cross-sectional, survey and administrative, public and private. Notes on data issues appear throughout this report, particularly in Sections 2.4. and 3.6.

Source of payment data is another important accounting topic raised by Jorgenson. The national accounts view GDP as an aggregate measure of economic activity that involves multiple sectors. In the case of medical care, households supply the patients; the business sector includes the providers of medical care; government plays an important role as a payer of a large portion of this medical care or, in some cases, by functioning as an insurer. Jorgenson, along with participants from CMS, noted the importance of keeping track of payments for medical care by the various sources.

As noted above, the national health expenditure accounts compiled by CMS are organized by sources of payment—the part paid for by the government, the part paid for by private insurance, the part paid out-of-pocket by the household sector, and so forth. Because the national accounts are used to monitor the government budget, they have to be able to indicate precisely the level of public expenditures used to purchase or pay in part for medical services. For private sources of payment, even though they constitute a relatively small portion of total payments, it is also important to distinguish between the parts made by individuals and by businesses. Here, the concern is not with the business of providing medical service, but the role of businesses as buyers of health insurance; the data must allow users to distinguish the health insurance industry and its activities from those of the medical care sector itself.

When thinking about payments for treatments generated by providers, Jorgenson noted that those have to be segregated, as they are in the national health expenditure accounts, by sources of payment, so that links can be made to the other sectors in the economy. These accounts support a key function, which is to document the flow of payments among households, industry, and government budgets. In developing satellite accounts, it is important not to lose sight of the tremendous heterogeneity that characterizes medical care and how that is reflected in part in the sources of payment data.

2

Allocating Nominal Expenditures on Medical Care: A Disease-Based Conceptual Approach

In the Bureau of Economic Analysis's (BEA's) phased plan for implementing a satellite account, the first major task is to define expenditure categories and devise a method for allocating economy-wide spending on medical care into those categories. Expenditure data are needed for multiple purposes—for health program administration, for the production of price indexes, for productivity analysis of the economy's medical care sector, for national income and product accounting, for disease treatment monitoring, and for making cross-country comparisons of health systems—and the ideal data set characteristics and organizational framework will be different for each.

Currently, the U.S. health expenditure accounts, produced by the Centers for Medicare and Medicaid Services (CMS), essentially track the flow of funds based on final payments from payers (private insurance, government programs, out-of-pocket) to payees (hospitals, physicians, drug vendors, nursing homes).[1] For many of the purposes raised during the workshop, the capability is needed to aggregate expenditure data into units defined along different lines—specifically, the real outputs of medical care. As Dale Jorgenson put it, the main objective is to collect data on the prices and quantities associated with the output of the sector and to cope with the enormous heterogeneity and the very rapid evolution of the character of the products, which differ both within and across providers. Workshop participants agreed that, because they serve as a building block for many kinds of health data systems, creating new ways of organizing and tracking health care expenditures is an immediate priority. This work would be useful for both

---

[1]To get a sense of the breadth of expenditure information produced by CMS, see the data tables produced on the agency's website (http://www.cms.hhs.gov/NationalHealthExpendData/downloads/tables.pdf).

the experimental health accounting and national income and product accounting purposes, as well as for price and productivity measurement.

Ideally, medical sector goods and services would be defined in such a way that: (1) expenditures could be estimated each period for every good or service produced by the industry, (2) meaningful quantities and prices (nominal and real) could be tracked, and (3) quality change of the goods and services could be monitored. The first task in the accounting exercise is to allocate nominal expenditures to the various array of outputs. Assuming that patients seek medical care to treat specific conditions or diseases, the medical care output should be defined and arrayed to reflect that consumption objective. Many of the researchers present at the workshop favor an episodes-of-treatment organizing principle for doing this.

2.1. METHODS FOR ATTRIBUTING SPENDING TO TREATMENTS: THE BIG COMORBIDITY ISSUE

During her presentation, Ana Aizcorbe identified several options for attributing spending across treatment episodes, or "disease buckets," as several participants described them. One is an encounter-based method in which spending is attributed to one or to several diagnoses as reflected by data extracted from patient claims. A second, broader approach involves constructing episodes of treatment—which may include numerous encounters over a predefined period—then adding up dollars spent nationally on each of the range of diseases and conditions. A third possibility is a person-based approach, in which spending on various treatments is tracked on a person-by-person basis over a set period of time.

Within these approaches, there are different techniques available for assigning the dollars spent to the treatment categories. The applicability and appropriateness of the methods varies by the accounting objective, and each has its pros and cons. Aizcorbe conceded that, at this point, it is unclear which is the best way to move forward for BEA's specific application. BEA is working both internally and with the Cutler-Rosen team to establish what the allocations may look like under the different methods, and whether it matters for estimating expenditures and prices (see Section 2.3.). Speaking about this project, which has begun producing episodes-of-treatment cost estimates, Allison Rosen noted that spending could also be further broken down into subcategories along functional lines, such as disease prevention, diagnosis, and screening activities. This is important, since not all spending on medical care can be attributed specifically to the treatment of a disease or condition.

Whichever method of allocating expenditures is used, it has to offer a solution to the comorbidity problem. Dealing with the reality that many patients utilize medical services for multiple conditions is a major issue to be confronted in health accounting. It is a problem on the expenditure side—BEA must figure out how to allocate spending for cases in which patients receive medical care for more than one disease. It is also a problem on the outcome side—how can researchers determine which treatments are incrementally affecting the quantity and quality of life among populations receiving care for multiple conditions.

Sherry Glied identified another dimension, distinguishing between horizontal comorbidity—multiple things happening to a patient simultaneously, which muddies the question of primary diagnosis—and vertical comorbidity—which deals with sequences of risk factors and also complicates designation of an episode of treatment. On one hand, if a patient's cholesterol is treated and that person never goes on to develop heart disease, how is that handled? Where are those expenditures grouped? On the other hand, if a person is treated for heart disease and is made even sicker because the drugs have side effects, where do those treatments fit? Are they part of heart disease treatment, or should they be categorized elsewhere? Rosen cited cardiovascular disease as perhaps the classic example of comorbidities. She noted that the place where comorbidity issues are most marked is in the risk factors—it is rare to see one factor without at least one other—and one could consider making separate buckets for patients in this group. For example, diabetes probably needs to be separated out because of all of the other complications that it causes. Glied concluded that, in many cases, defining what is a final product of the medical care industry is going to be a tough task for BEA to handle.

Encounter-Based Approach

One relatively simple method for reporting the cost of illness by disease involves tracking spending that takes place at the patient encounter level. Information about the cost of specific patient encounters with the medical care system can be found in administrative data, such as claims forms (often, these have the payment the provider requests, but the payer pays something less than that); expenditures can be allocated based on diagnosis identifiers. The Altarum research (described below) used primary or first-listed diagnostic category for this purpose. Within this method, disease buckets can be allocated at varying levels of detail. For the Altarum project, 660 clinical classification categories were used based on groupings created at the Agency for Healthcare Research and Quality (AHRQ). Figures for these categories can, if desired, be aggregated into a smaller number of buckets.

Aizcorbe made the point that relying on the primary diagnosis may seem like a coarse decision rule, but in fact these kinds of compromises are often needed in the production of the national accounts. Firms (or even establishments) exist that produce a range of different goods, sometimes across more than one primary industry, and their outputs have to be allocated. In such cases, BEA analysts must figure out where to allocate dollars associated with each type of output in the industry accounts. It is important to have time series data produced on a consistent basis, even if the way that the dollars are allocated is not completely accurate.

Rosen pointed out that there is a rich history of using encounter-type approaches in cost-of-illness estimation for the health care system, going back to Rice, Hodgson, and others (Rice, 1966; Rice, Hodgson, and Kopstein, 1985); a number of these studies have used data from the Medical Expenditure Panel Survey (MEPS). Among the pros of the approach is that calculating spending and attributing it to diseases is easier on an encounter basis than it is for some other kinds of measurement units. On the downside, the encounter approach is fairly limited in its capacity to handle comorbidities; a number of providers do not use claims (though some organizations, such as Kaiser and the Veterans Administration, do have, in effect, dummy claims that associate costs with services) or use claims that do not always provide valid disease diagnoses.

Episode-Based Approach

An alternative organizing principle is a medical care or treatment episode, which is a broader concept than an encounter. As Rosen explained, under this approach, claims are organized into clinically distinct episodes of care that are "adjusted for disease severity and complexity." In the case of a heart attack, the episode involves not only a patient's hospital stay, but also the convalescent time and the care that is given afterward over some discrete window of time. In addition, because there might be an acute myocardial infarction or an acute myocardial infarction complicated by congestive heart failure, there can be varying degrees of severity within a given disease; ideally this variation would be accounted for in the expenditure allocation.

A number of commercial firms create so-called episode groupers. As described by Rosen, a major purpose of the groupers is to try to get at some of the differences between chronic long-term diseases and acute short-term diseases. For example, for the office visit of a diabetes patient, the doctor may assign an International Classification of Diseases (ICD-9) code, and the spending will be attributed as such under an encounter-based approach. However, the doctor may also give a prescription for a medication for hypertension, at least in part because of the patient's diabetes. If that is not something that gets picked up on the pharmaceutical claim, then it may be assigned to another category. The grouper methods attempt to identify a window of time so that, for the diagnosis of diabetes, all spending for some predetermined length of time would be assigned to that disease. For another diagnosis, the appropriate time period may be different. Using the disease classification codes (such as ICD-9) to categorize patients is fairly straightforward; what needs considerable work is the question of determining the rule set for defining what that chronic episode looks like and how long it lasts.

The episode unit of analysis can be distinguished from the encounter-based approach in that it consists of groups of claims that take place over an expanded, variable time window. These characteristics allow users of the approach—who already include a number of commercial insurers—to take into account both the per-unit cost and the volume and mix of services. At this point, the major problem is that grouper software is proprietary, and the algorithms underlying these tools are concealed. Ideally, a statistical agency's products should embody methodological transparency to users.

Person-Based Approach

The person-based approach estimates expenditures as a function of spending at a different unit of analysis—the individual who is being treated for some vector of diseases or conditions. In their broad-based health accounting work, Cutler and Rosen have been using regression techniques to assign spending across disease episode treatments at the person level. The dependent variable is cost, or total expenditures on medical care, which is regressed against a set of disease dummy variables. The expenditure period is a prespecified time window—Cutler and Rosen have been using one year. The results provide a picture of the incremental per-patient annual spending attributable to each disease category.

Rosen expressed the view that the person-based regression approach is probably the best of the options for handling comorbidities. At the event or encounter level, many patient contacts with the medical profession (doctor visits) are attributed to a single specific reason, if they are coded at all. Although there can be a huge number of disease buckets (Aizcorbe noted about 700,000 if combination categories are allowed in the claims data), clinical knowledge can be used to narrow these down to essential groups of comorbidities. For example, the cardiovascular disease risk factors might include hypertension, hyperlipidemia, coronary heart disease, and the like. Given the limited sample sizes of available data sets such as MEPS, it is absolutely necessary to identify the most relevant comorbidity combinations.

During open discussion, Joseph Newhouse made the point that, since no accounting system can manage 700,000 separate disease buckets, some will be collapsed into the regression's residual category. Implicitly the magnitude of this residual is dictated by how many interactions are specified in the regression. He wondered whether, for the comorbidity issue, the magnitude of the problem was more or less the same in the episode and the regression approach, because it all turns on what is specified in the interaction terms.

Another attractive conceptual feature of the person-based regression approach is that it can be readily matched to health improvements, because analyses on the health services and outcomes side tend to be monitored patient by patient. Health improvements are not typically monitored immediately before and after treatment; researchers look at how a person or group of persons fares over some period of time after receiving treatment. The costs of cases for which there are no valid claims or ICD-9 codes can still be attributed through surveys as well, which is another positive feature.

These features notwithstanding, the person-based regression approach does have limitations. Ralph Bradley of BEA raised several methodological issues that have to be confronted, particularly if the method were to be used in price index construction. One has to do with population sample coverage; if an analyst runs the regression using data for the entire population, it will yield one set of coefficients; if data are separated, for example, between those who are insured and uninsured, an analyst will get completely different sets of coefficients. Alan Garber (Stanford University) agreed with Bradley that the analysis will be heavily sample dependent. In order to minimize the omitted variable bias, a representative sample is required. Any causal interpretation of the disease dummy coefficients would be problematic. If—in trying to answer a question like, "What would happen if we eliminate a disease by using a particular drug or doing a particular operation?"—an analyst simply plugs results into a model to estimate overall expenditures or drug expenditures, it is likely to be wrong, because there will be omitted variables and the change will not be the same as predicted by the sample from which the data were generated. There is also the issue of how, in the regression approach, to allocate the intercepts for the base spending for the year. These criticisms aside, Garber expressed sympathy for the regression approach, in part because there are not many alternatives. The key is to be cautious about how the model's coefficients are interpreted and applied.

Rosen agreed that the approach has problems to be overcome, and they tend to be related to the tremendous amount of heterogeneity in terms of who gets what medical care. Matthew Shapiro made the point that, ideally, one would want to stratify the results. For example, the elderly will have a very different spending profile for certain diseases than the young, and one would like to be able to deal with that. With advanced age, comorbidity is much more likely to be present. Therefore, adding up simple cases—diabetes, heart attack, etc.—is not going to work very well; it might be hard to extrapolate from a 50-year-old's noncomplex heart attack to what would happen with a 70-year-old. Fortunately, data exist with which to investigate these issues; however, the more that the analysis is driven to define activity at group levels, the greater the required sample size becomes. Shapiro added the related observation that, if database size were not an issue, one could think of comorbidities as separate diagnoses; a simple heart attack would be one diagnosis, a complex heart attack or heart attack plus diabetes another.

As discussed in Section 2.3., which cost allocation method is best will differ on the basis of how it is to be used. For the creation of price indices, a person-based approach may not be as appropriate as an episode-based approach. If the goal is to broadly relate cost and health improvements or to compare costs and health improvements within a given disease on a micro level, as done in cost-effectiveness studies and decision analysis, that might be better done with a person-based regression approach.

2.2. ALLOCATING PERSONAL HEALTH EXPENDITURES BY MEDICAL CONDITION: THE ALTARUM INSTITUTE PROJECT

Charles Roehrig presented his and his colleagues' work at Altarum, a nonprofit institute based in Ann Arbor, Michigan, on reallocating national health expenditure (NHE) estimates produced by CMS into medical condition categories. This work evolved from efforts by the institute to develop a model to forecast national health expenditures consistent with the NHE accounts. Their interest was primarily to further understand the drivers of health care spending growth and the prevalence of medical conditions. The project, a nine-month effort supported by the Pharmaceutical Research and Manufacturers of America, benefited from the advice of several experts on measuring health care expenditures including Linda Bilheimer, Mike Chernew, Joel Cohen, Mark Freeland, Rod Hayward, Steve Heffler, and Judy Lave—several of whom attended the workshop.

Roehrig detailed through how the project allocated expenditures by medical condition. The first step involved revising the NHE revenue categories to create a more function-oriented picture. For example, hospital-owned nursing home revenues were shifted from the "hospital" category to the "nursing home" category. The "purified" service categories consisted of the following:

- Hospital
- Physician
- Prescription drugs
- Nursing home
- Home health
- Dental
- Other professional
- Other personal
- Durable medical equipment
- Nondurables

For reasons that become apparent below, MEPS records also had to be mapped into the NHE categories, as shown in Table 2.1.

The method for reallocating expenditures into the purified categories was based primarily on a detailed study done jointly with AHRQ and the Office of the Actuary at CMS. The results for year 2002 are shown in Table 2.2. The first column shows how the $1.3 trillion in personal health expenditures were allocated by the original NHE service types. The post-reallocation numbers, intended to provide a more functional picture, are listed in the column on the right. Of note is the large reallocation of expenditures out of the hospital (8.1 percent) and physician (15.6 percent) categories to the others (for example, 2.7 percent to home health and 4.4 percent to nursing homes). The totals under the two structures are the same but, according to Roehrig, the hospital expenditure is more closely aligned with what most of us think of as hospital services.

TABLE 2.1 Mapping of MEPS Event Categories into NHE Service Types

MEPS Event Type	Charge Type	NHE Service
Inpatient	Separately billed doctor (SBD)	Physician
Inpatient	Facility	Hospital
Outpatient	SBD	Physician
Outpatient	Facility	Hospital
Emergency Room	SBD	Physician
Emergency Room	Facility	Hospital
Office based	Doctor	Physician
Office based	Other provider	Other services
Home health	n/a	Home health
Prescription drugs	n/a	Prescription drugs

NOTE: n/a = not available.
SOURCE: Workshop presentation by Charles Roehrig.

TABLE 2.2 Sample Calculation of Medical Care Expenditures by Functional Category for 2002 (in billions of dollars)

Service Type	Baseline NHE	Shifts Out	Shifts In	Purified NHE
Hospital	488.6	39.6	0.0	449.0
Physician and clinical	337.9	52.7	0.0	285.2
Dental	73.3	0.0	0.0	73.3
Other professional	45.7	1.8	33.7	77.6
Home health	34.3	5.7	13.3	14.9
Nondurable medical products	30.9	0.0	0.0	30.9
Prescription drugs	157.9	0.0	10.1	168.0
Durable medical equipment	20.8	0.0	8.2	29.0
Nursing home	105.7	0.0	21.3	127.0
Other personal care	46.3	0.0	13.2	59.5
Total	1341.4	99.8	99.8	1314.4

SOURCE: Workshop presentation by Charles Roehrig.

The second step of the allocation exercise was to calculate the distribution within each functional expenditure category by population group; the designated groups are the civilian noninstitutionalized population, various institutionalized populations, and active-duty military, because that is how the data sources break down, more or less. The researchers primarily used the MEPS-sourced data developed by Sing et al. (2006).

Finally, for each functional category by subpopulation cell, expenditure totals were distributed by medical condition. Altarum used the AHRQ clinical classification system, which Cutler and Rosen have also used in their project. The civilian noninstitutionalized population accounts for the overwhelming share of spending. For example (again based on analysis of the Sing et al. data), 84.2 percent of hospital spending was by the civilian noninstitutionalized population; the next highest population group—nursing home patients who had an acute episode and were admitted to the hospital for a period—accounted for a comparatively modest 6.5 percent. Similarly, about 82 percent of personal health expenditures were attributable to the civilian noninstitutionalized population, in the sense captured in MEPS, and another 14 percent to the nursing home population.

Altarum relied heavily on MEPS for data on the civilian noninstitutionalized population. MEPS provides spending by person, encounter or event, type of service, and the medical condition broken down into 260 Clinical Classification Software (CCS) categories by 7 service types—in some instances, multiple conditions are present, and sometimes there are missing conditions. For care delivered to nursing home residents, the researchers used data from the National Nursing Home Survey. For nursing home residents who were admitted to a hospital, Healthcare Cost and Utilization Project data were used. Roehrig reported that they plan to use the Medicare Current Beneficiary Survey for future work. The project's final database includes 10 years of data from 1996 through 2005, all of the years for which MEPS data are available.

Altarum was unable to attain conditional distribution information for about 9 percent of personal health expenditures on items like other nondurables (e.g., tissues, things bought at the pharmacy) and durable medical equipment. Roehrig indicated that they would be able to allocate these items. "Other personal care," a catchall category, includes such items as industrial implant services and Medicaid waiver programs aimed at keeping people in their homes and out of nursing homes that are difficult to assign to specific categories.

Next, Roehrig explained how they dealt with comorbidity—patients in the MEPS data set with multiple conditions. The vast majority of expenditure data in MEPS is on individual events—inpatient episodes, outpatient visits, prescription drugs—that have only one condition assigned to them. A patient could have multiple medical issues but, for example, if he breaks a leg, the treatment record typically indicates just that primary purpose. At the event level, the issue of comorbidities is not nearly as conspicuous as it would be in a person-level or even episode-level analysis.

Roehrig also noted that there are sharp differences across medical conditions—some show up much more often with comorbidities. For example, inpatient events for back problems almost always show up in MEPS with that singular condition; the same is true for cancer. However, inpatients hospitalized with diabetes or hypertension more often than not have other conditions recorded in the MEPS data.

The project team considered a couple of ways of dealing with these comorbidity problems. The simplest option is an unweighted allocation—if the patient has two conditions, spending is split 50-50 between them; if there are three conditions, it is split in thirds. Roehrig termed this the proportional approach. The second option is a weighted approach that takes into account the average cost of an event for each condition when is appears alone. So, for example, the estimate for diabetes is based on the average cost of all hospital events that are just for diabetes. The same is done across all conditions. If diabetes appears with a second condition, the two are weighted proportionately with their average stand-alone costs. This is the approach that was ultimately used for the research effort, largely on the grounds that it made intuitive sense. The exception is the nursing home population, a great majority of whom display multiple conditions. Here, the weighted approach was not feasible and the unweighted allocation was used.

Roehrig then presented the study's results. Figure 2.1 shows expenditures by diagnostic category over a 10-year period; 262 AHRQ CCS categories were grouped into ICD-9 codes. The circulatory system category accounted for the largest share, about 17 percent of personal health expenditures. The next seven codes each contributed between 6 and 9 percent of the total. Approximately 50 percent of expenditures are captured in these groupings.

MEPS data were also tabulated to estimate the most costly medical conditions. (See Table 2.3.) Comparing these levels with those from 1996 allowed Altarum to estimate spending growth rates for medical conditions. Pneumonia, chronic obstructive pulmonary disease, lung cancer, stroke, and coronary heart disease were categories showing the slowest expenditure growth rates, all at 4 percent or less. This may reflect some beneficial effects of reductions in smoking over the period.

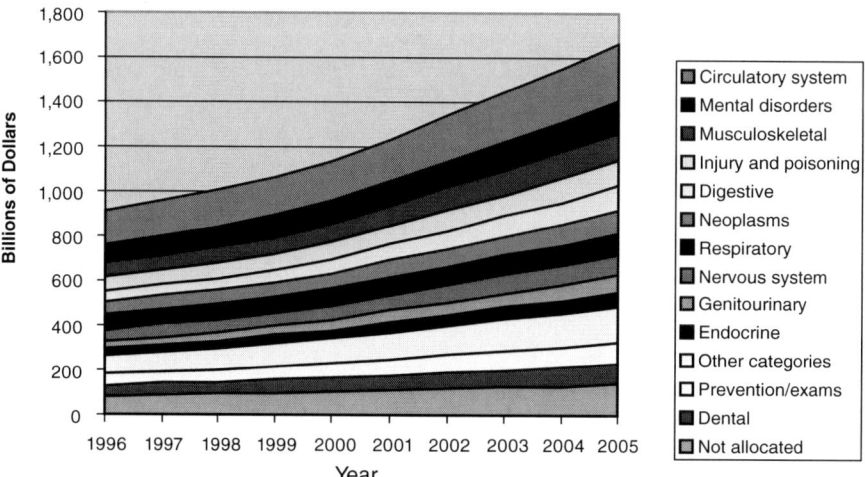

FIGURE 2.1 Annual health care expenditures by diagnostic category.
SOURCE: Workshop presentation by Charles Roehrig.

TABLE 2.3 The 15 Most Costly Medical Conditions, in Terms of Personal Health Expenditures (PHE), 2005 (in billions of dollars)

Medical Condition	PHE
Mental disorder	142.2
Heart conditions	123.1
Trauma	100.2
Cancer	99.4
Pulmonary conditions	64.6
Hypertension	50.2
Osteoarthritis	48.0
Back problems	40.1
Kidney disease	35.9
Diabetes	35.8
Endocrine disorders[a]	29.2
Skin disorders	27.2
Cerebrovascular disease	26.8
Hyperlipidemia	22.8
Infectious diseases	22.5

[a]Excludes diabetes and hyperlipidemia.
SOURCE: Workshop presentation by Charles Roehrig.

2.3. COMPARING THE METHODS

Over the past year or two, the Cutler-Rosen group has been working with BEA to empirically assess differences in the various approaches to allocating medical care expenditures by disease; Rosen reported some preliminary findings. The research objective is to reconcile disease categories among the encounter-, episode-, and person-based regression approaches; to simulate costs of diseases using each; and to compare and contrast the findings. For this project, Rosen and Cutler have been using health claims data for the period 2003-2005 from Pharmetrics Inc. For 2003, the data cover just over 3 million patients and include total spending of $9.09 billion on inpatient and outpatient services, office visits, prescription drugs, skilled nursing facilities, and laboratory services. Up to four ICD-9 diagnoses are present on a given claim, although only the primary diagnosis is listed for hospital claims. Symmetry software from Ingenix was used to link medical expenditures to disease categories.

In order to reconcile the three approaches to common disease categories, Rosen et al. first mapped ICD-9 codes into CCS categories. These were aggregated into 65 clinically meaningful groups that had been developed earlier based on advice from physicians. Cost categories were created primarily for diseases with known treatments that have led to health benefits and for which more detailed analyses could be done matching quality to costs.

For the person-based regression approach, the authors were able to use all listed diagnoses on claims in a given year. For the encounter approach, an algorithm was used to determine the diagnosis (usually the first listed) to which the majority of spending went, and the dollars were assigned to that category. For the episode-based approach, each episode treatment group (ETG) was allocated to the clinical group that accounted for the largest share of spending. There were a few problems with the ETG approach. For example, there was no ETG for cervical cancer; also, the transparency issue remained—the method of aggregating data into the ETGs was still essentially a black box.

Under the encounter approach, about 19 percent of the spending recorded from claims had no listed diagnosis, and the dollars could not be allocated to a condition. Using the episode-based approach, only 1 percent of spending originated from claims with no ETG. Using the person-based approach, expenditures for individuals with no diagnoses accounted for only about 0.6 percent of the total. The problem of unlinkable spending was clearly most serious with the encounter-based approach.

The Rosen-Cutler work demonstrates that the cost of illness can be estimated by each of the proposed methods. The total dollars that can be allocated differs, and, certainly, the fact that noncomparable data sets are being used for the different methods also has an impact on the results. Table 2.4 shows annual spending by condition estimates. The spending estimates varied significantly for some disease categories. For example, the person-based approach yielded very high annual expenditures for dementia—on the order of $9,000—relative to the encounter- and the episode-based approaches. The likely reason is that the regression used does not include all of the needed interaction terms, so the estimates essentially capture unobserved correlates of spending. Instead of getting just spending on dementia, the coefficient is picking up aspiration pneumonias, feeding tube treatment, and other things for which clinically meaningful buckets need to be created. This, Rosen emphasized, is why it is important to bring clinical insight into analysis.

For some of the same disease categories, the encounter-based approach appears to underestimate expenditures. Rosen noted that this may have something to do with the way risk factors for diseases are commonly coded. For example, physicians may be more likely to code coronary heart disease than they are diabetes, hypertension, or hyperlipidemia. In contrast to the encounter-based approach, which relies entirely on physician coding on claims, one nice feature of the person-based approach is that coding can be captured over time, so more information about multiple conditions can be obtained; the approach also allows the claims data to be supplemented with surveys, injecting information from patients that can enrich the picture.

Rosen also noted that they have not yet done any time series using MEPS data. Making some direct comparisons, they have found some of the same things—for example, dementia and acute renal failure, which tend to occur in patients being treated for other conditions simultaneously, end up being much higher with the person-based regression approach than the others.

TABLE 2.4 Annual Per-Person Cost for Selected Diseases by Method, 2003

Disease	Encounter	Episode	Person
Colon cancer	$8,100	$4,458	$10,475
Lung cancer	12,082	14,213	23,895
Dementia	596	1,111	9,231
Depression and bipolar disease	616	984	1,070
Hypertension	225	522	376
Coronary atherosclerosis disease	3,415	4,342	3,303
Congestive heart failure	2,869	2,476	12,645
Cerebrovascular disease	2,563	2,818	5,759
Asthma	348	639	519
Chronic renal failure	11,105	11,433	11,964
Osteoarthritis	1,184	1,726	1,450

SOURCE: Workshop presentation by Allison Rosen.

At this point, criteria for ranking the suitability and accuracy of the different methods have not been fully sorted out; Rosen noted that the right answer has a lot to do with the specific applications. She also encouraged others in the community to provide feedback on the topic. If the best unit of measurement is the episode, whether defined by an existing grouper or in some other way, it is important to proceed so that the underlying approach that would be used by government agencies is transparent.

During open discussion of the different methods for allocating expenditures across conditions, Steven Cohen (AHRQ) pointed out that the Centers for Disease Control and Prevention, AHRQ, the National Pharmaceutical Research Council, and the Medicare chronic disease directors have developed a chronic disease cross-calculator for the attribution of costs across different conditions. He applauded the work by the Cutler-Rosen group and others at the workshop to help understand the nuances in terms of the different methodologies.

Cohen noted that the opportunity to look at distributional aspects in the costs of chronic illness treatment is incredibly useful to his agency; having information about the concentration of expenditures and the characteristics of treatments will be particularly applicable for estimating the impact of preventing or reducing the incidence of some of these conditions—both in terms of valuing health care and monitoring potential savings. The metric might not be cost savings, but better value for the dollar. Mark Freeland (CMS) added that it was exceedingly valuable for his agency to see how different the results can be and that there is some purpose—price index construction, national benefit cost analysis, the national income and product accounts, etc.—for which each of the constructs may be the best.

Likewise, Jack Triplett was encouraged by the work to improve data and methods for allocating nominal medical care expenditures, noting the large amount of progress that has been made over the past 10-20 years. In the past, efforts to disaggregate the national health accounts into disease categories have not produced any time series data. When Triplett estimated a real expenditure account for mental health a decade ago, an enormous amount of work was necessary to reconcile an existing series of annual cross-section estimates because, in their construction, no attention had been paid to making new estimates compatible with those of earlier years. Now, not only is work progressing on time-series data, but also several alternative methods are being discussed; Triplett cited this as a big step forward.

Shapiro endorsed BEA's cautious approach. He suggested that the agency should probably not buy into a particular approach too early, adding that it would be useful to see whether the choice actually matters for the statistics. This means that there might need to be parallel sets of accounts going on, at least on a research basis, for some time.

2.4. DATA NEEDS FOR EXPENDITURE ACCOUNTING

Workshop participants touched on data needs at many points during the day's proceedings, and that discussion is sprinkled throughout this report. This section summarizes a few of the key data themes that emerged.

Drawing from Multiple Sources

Dale Jorgenson emphasized the need to consider a wide range of possible data sources on medical treatments to underlie the satellite account. The country's health system is characterized by a lot of patient and treatment heterogeneity, and it is not easy to collect this information, especially at the level of detail needed to capture the rapid evolution of the character of medical treatments. For measuring the quantity and price of these treatments, one can look to providers, customers, and third-party payers.

Service provider data are the starting point for the producer price index at the Bureau of Labor Statistics (BLS), and Jorgenson expressed the view that this type of information is going to have to play a role for the BEA work, particularly on the industry side of the accounts. Another source from which information can be collected is the customer. Jorgenson noted that this is where the distinctive features of the medical system really come into play. Unlike some other areas of the economy, it is difficult to get accurate and relevant information from the consumer (patient). Patients pay a relatively modest portion—some information can be collected from the patient about the episode and the treatment, but when dollar figures are needed, one has to go to the provider. The final source is the third-party payer, which plays a big role in the strategy laid out by BEA. A vast amount of claims data are maintained by both private and public payers.

Ideally, these data would be matched against the payments that are received by the providers.

Jorgenson commented that he did not think it would be fruitful to look at the data issue as an either/or proposition. In the medical panel data, the MEPS survey, an attempt has been made to combine the information collected from providers with information collected from payers. This, he said, is a good practical approach and one that should be encouraged. His recommendation to BEA and BLS is that they should coordinate their efforts to get the most reliable information on prices with data from the payers (i.e., claims) and also from the providers. In thinking about measuring prices and quantities, the relevant data that can be brought to bear will have to include the sort that the producer price index already collects.

Aizcorbe agreed that all data sources that could usefully supply the accounting system should be considered. She did, however, raise one deterrent to extensively using data from providers, which is how to link expenditures to patients. Using the example of depression, she questioned how the different care elements—the doctor visit, the drug purchase, and the talk therapy—could be linked together for the same patient. So far, in the satellite program, BEA has not used the raw data that BLS uses for its indexes exactly for this reason—they cannot be linked to patients. The MEPS data do link to patients, which is what BEA needs for at least a sample of the population.

For most of the data approaches being considered by BEA, the treatment is what needs to be priced; this requires data at the patient level reflecting the full combination of inputs into the treatment. Jorgenson made the point that a solution will still need to be found for combining data on providers' prices with the information collected from claims, and he suggested that kind of work be put on the table for BEA. Aizcorbe agreed, remarking that it is important for the agency to think about what it will be doing 10 years from now; it is not obvious yet how to take the data that underlie the price programs at BLS and use them directly for BEA's purposes. Another aspect of the data coordination task involves reconciling the microdata in MEPS with the national health expenditure accounts, because they don't add up to the same national totals. That is primarily because the scope of the populations and of the spending are not quite the same, a situation that calls for regular updating of the reconciliation work done by AHRQ and CMS. Aizcorbe reiterated the importance of working out how to coordinate and exploit multiple data sources, as well as anticipating how this strategy will play out once the research program is in full swing.

Sample Size: Capturing Information on Less Common Conditions and Morbidities

David Cutler pointed out that, when relying on survey data sources such as MEPS, the key challenge is that they do not include enough patients to capture rare diseases or less common comorbidity combinations. For example, not many lung transplants are going to appear in MEPS, which in 2005 had a sample size of 15,000 families and 39,000 individuals. Cutler suggested that 30 million person records would be needed to cover the full range of diseases and combinations thereof. Since such a massive expansion of survey sources is impractical, a next best option may be Steven Cohen's suggestion to oversample rare diseases.[2] Several workshop participants also suggested that, ideally, any results should be stratified to account for different spending profiles by age for certain diseases.

Regarding other population coverage gaps, Constance Citro made the point that the American Community Survey may offer an opportunity to cover some institutionalized populations (through its group quarter sample) omitted from the scope of MEPS. Cohen added that CMS conducts the Medicare Current Beneficiary Survey, which captures a fairly large segment of the institutionalized population. He reported that his department's data council has been thinking broadly about where the gaps are for other individuals in long-term facilities.

Aizcorbe acknowledged the value of these ideas, noting that BEA (as well as Cutler-Rosen) are looking into some of them already. A lot could be done with claims data for the insured population simply because of their enormous size and coverage. BEA participants agreed that using MEPS as the backbone of the data infrastructure, and then claims information in a supplemental role wherever gaps appear, was a reasonable strategy. Even so, treating different comorbidities as separate disease categories—which Aizcorbe agreed was a good idea and would have to be done to some (as now unknown) extent—still runs up against data inadequacies. Even with the largest data sources, a portion of spending occurs in buckets that have very few observations, and creating separate categories for comorbidities still does not always work. This is why medical expertise is needed when setting up the account structure.

Data Representativeness

The tradeoff between sample size and representativeness is one data issue to which workshop participants returned on several occasions. The work by the Cutler-Rosen group has relied heavily on microdata from national surveys that sample individuals, such as MEPS, supplemented with the Medicare Current Beneficiary Survey. As noted above, while the survey data are essential to the accounting project—and very useful for high-prevalence conditions, particularly the cardiovascular disease and cardiovascular disease risk factors—there are real sample size inadequacies for conditions with lower prevalence. In contrast, the insurance claims data provide a large sample, but at the expense of representativeness—no single source provides a national sample. It is easiest to

---

[2]This then begs the question of how to find these people; MEPS uses a sampling frame built from the National Health Information Survey, which records only self-reported diseases.

find data on people in large firms with standard kinds of benefits. However, there are no large samples of uninsured people; for this group, something like MEPS will inevitably have to be more heavily relied on. Newhouse added that the transaction prices for the uninsured are also very complicated. The hospital trying to collect on a debt may turn it over to a collection agency and may agree to some payout period that stretches over months if not years.

Another problem with existing claims data sources, which BEA is struggling to get a handle on, is that they typically track patients only as long as they are covered by a particular plan. So changes in employment could lead to discontinuities in the data. If a person switches jobs—and even if both the old and new insurance plans are in BEA's database—it may not be possible to connect the records in a way that ensures that one is dealing with the same patient. This may not be particularly worrisome if changing jobs or plans is not highly correlated with disease incidence or conditions. However, Aizcorbe pointed out instances for which that may not be the case. People may select less expensive plans, such as health maintenance organizations (HMOs), until their situation changes; for example, if a woman becomes pregnant, who the plan provider is may suddenly become important, and she may switch to a different kind of plan.

At this point, it is difficult to know exactly what the optimal balance will be in terms of how to utilize the different kinds of data. It is clear, however, that the satellite accounts will need to draw from many data sources and methods will need to be developed to coordinate them. Combining the survey sources with Medicare records can reach a significant share of the population, but it is unlikely that data will be comprehensive to the point of providing a picture for a group of the population that is completely random any time soon. Commenting that systematically missing data coverage is especially worrisome, Aizcorbe noted that statistical methods can be used in the accounts to minimize some of these problems. For example, weighting is used to correct for the fact that the annual Survey of Manufacturers disproportionately samples large firms (those with more than 5,000 employees). She also noted that, even if there were enough of every different type of patient and plan in the data, reweighting would be needed to make it align with the sampling frame of MEPS or some other national surveys.

Cohen reaffirmed comments by Cutler and Aizcorbe about the need to ensure the national representation of data sources. He pointed out that, for many disease areas, expenditures are highly concentrated. Some of the uninsured, for example, are in that predicament because they have chronic diseases; as a result, data could be highly skewed in terms of the segment that is missing. He added that communication among BLS, BEA, and the Department of Health and Human Services is essential as the agencies think long term about needs and potential oversampling strategies to fill gaps in a much more efficient manner. Given the limitation of departmental resources for surveys, Cohen noted the importance of opportunities to link MEPS to the National Health Interview Survey to create ways of predicting the likelihood of an individual being uninsured in the long term; subsequently, high probability portions of the population could be oversampled. He added that, if resources permitted, it may be possible to look at specific chronic diseases in which there are well-developed evidence-based processes of care. Even with fixed resources, there may be ways of differentially sampling the population in order to meet both departmental objectives and to help inform policies and programs at other agencies. Cohen reported that conversations have already begun taking place between BEA and BLS about how to help meet the needs that BEA has on the spending side versus the needs that arise on the industry side.

Looking down the road, the question of how big a hindrance to health accounting the lack of data representativeness will be is a major one. For research purposes, if partial pictures can usefully be explored, it is less of a problem. For the national accounts, which must be complete and national in scope, the problem is more severe and may require short-term compromises. For something like measuring quality change of treatments (discussed in Chapter 3), the satellite account methodology may have to rely on inferences based on more common diseases, at least for a while; this would seem better than no quality adjustment at all.

3

Price Indexes: Calculating Real Medical Care GDP

Although the term "satellite health care account" features in the title of the workshop, much of what was discussed over the course of the day had to do with medical care price deflation—the use of price indexes to estimate real changes in the levels of inputs and outputs for the sector. Matthew Shapiro, who has done seminal work on the topic, began his comments by noting that a big part of the task for the Bureau of Economic Analysis (BEA) involves parsing out nominal expenditures in a way that is meaningful and conducive to measuring prices.

BEA is already in the business of developing price indexes for the purpose of calculating real levels of economic activity, on an industry by industry basis, for the national accounts; this responsibility is particularly demanding for the medical sector in which third-party payments, and the fact that transactions do not occur in textbook competitive markets, confound price measurement. Some aspects of this problem have already been dealt with on the nominal side of the accounts by allocating the actual expenditures back to the consumer, to the government, and so on, regardless of who actually pays, which is often an insurance company.

Once the nominal flows for the sector have been figured out correctly, which is a daunting problem in itself, the next task in developing the new BEA health care account is to begin estimating the disease-based price indexes. Shapiro endorsed this two-stage strategy, although he noted that, for other purposes, there were different ways—in addition to the disease unit concept—that are also useful health care price measurement. For example, hospitals would want to know about prices specifically for its industry. However, to get the price indexes from the consumer standpoint, disease by disease unit pricing seems more appropriate than a traditional industry-based approach or than a global pricing of population health. Even if, ultimately, the goal is to measure the price of an incremental gain in health, Shapiro argued that, for economic accounting purposes, one is driven by the logic of the disease by disease approach, which was the focus of much of the workshop.

A key question, raised by Steve Heffler (Centers for Medicare and Medicaid Services), is how the methods for parsing nominal expenditures by disease (described in Chapter 2) relate to appropriate price measurement. Ana Aizcorbe responded that those working on satellite accounts—the Cutler-Rosen group or BEA—first establish a number of disease "buckets" that make sense to the medical community; these buckets become the unit of observation for which spending and health effects data are collected. The dollar total spent per patient on a particular category—for example, diabetes—becomes the price for the newly defined unit of health service. Then, nominal spending on diabetes is given a weight based on its share of total medical care spending.[1] Likewise, indexes for each disease category are weighted then aggregated. Aizcorbe cautioned that BEA is still in the phase of attempting to figure out the best way to define diseases, and that developing these kinds of indexes is still a ways off. BEA has purchased some databases covering patients who are commercially insured with the intention of experimenting with different types of indexes and different ways of defining diseases.

In this context, Aizcorbe described the most important problem with producer price indexes for purposes of the satellite account envisioned by BEA: they do not identify the medical care good or service that is sought by the consumer—which most think should be the treatment of a particular disease or condition. She added that health economists have developed the conceptual tools that are needed to remedy the situation, and that putting these approaches into practice is something that BEA would be working on right away.

Much of the academic literature has relied on patient claims data to provide a picture of price trends for treating specific conditions. The economic good has been defined as a completed episode. For example, for a heart attack patient, this may involve time and expenditures on a series of initial treatments plus those that take place during the recovery period. At the end of that episode, data are collected to estimate all dollars spent over the entire period; this forms the basis for pricing a completed episode.

To develop a comprehensive health care account with this kind of underpinning, claims data would be needed for as much of the population as possible; Medicare and Medicaid would provide large portions. However, there are some groups for which claims data will not be available—most obvious are the uninsured, who do not submit claims—so their spending would have to be measured another way. Patients from some kinds of institutions are also not typically

---

[1] In a fully evolved price measurement program, tracking nominal dollars spent on treatments would be viewed as only a first step. As discussed in detail in Section 3.3., a fully meaningful price measure must ultimately also consider how the quality of a treatment changes over time.

included in these sources, so BEA will be investigating ways of obtaining spending data for them as well.

Aizcorbe also noted that the timing aspect of the treatment-based unit does not lend itself seamlessly to deflation in the national accounts. For example, for delivery of a baby born in January, most of the dollars are for services provided in the previous nine months. So, when pricing the completed episode, the reporting takes place in the year following the nominal expenditure outlays. Aizcorbe stated that, ideally, the price index should line up with the time period in which the spending actually occurred. There are other areas of the national income and product accounts (NIPAs) that share this issue (payment, consumption, or return from investment takes place beyond some point); one way to handle it is to think in terms of the price per patient over some predetermined period of time.

3.1. PRICING TREATMENTS TO CAPTURE CHANGING TECHNOLOGIES, INPUT SUBSTITUTION, AND POPULATION HETEROGENEITY

A disease-specific index must embed a capability to capture the substitution of medical care treatment inputs that takes place over time. Aizcorbe used the example of treatment for depression, which has transitioned from a high to a low reliance on talk therapy as less expensive alternatives—specifically antidepressant drugs—were introduced and proliferated. Tracking patients with this condition over the past few decades would have revealed some portion switching from talk therapy to drug therapy. As this has occurred, the average amount spent on treatment of depression has fallen. However, standard price indexes do not pick up this change because they track the price of talk therapy and of drug therapy independently, and therefore they do not catch the fact that people are switching from one to the other. Even if there is no innovation in prescription drugs and no price change in either approach to treating depression, the per-patient cost falls because this substitution has occurred. If the standard indexes are used to deflate nominal spending, the resulting measure ends up showing a drop in real spending or a drop in the quantity, when in fact the same number of patients are being treated for depression. For its satellite health care account, BEA proposes to take the system-wide spending over some period of time in a treatment (such as for depression), regardless of treatment mode, and divide it by the number of patients treated. The idea is to calculate a unit value that counts all of the spending and allows for substitution across treatment types for each specific condition.

Patricia Danzon, who spoke about the pharmaceutical industry, cited the growing prevalence in the market of biologics—biological products made from living organisms whose uses are similar to conventional drugs—as another example of a switch in technology that BLS will need to confront. At the moment, both a pharmaceutical index and a biotechnology index exist. But, Danzon noted, if the goal is to estimate change in pharmaceutical prices accurately, then the current movement toward more biologics and fewer chemical-based drugs needs to be captured. To the extent that they are measured in separate industries, a problem arises because the index will not capture the biologics that are dispensed through physicians' offices or retail pharmacies. They are likely to be picked up by Medicare as Part B drugs, but probably for many of the other databases, they are just part of physicians' services. As these biologics become a significant share of total pharmaceutical spending—and they will, especially at the expensive end—it will become increasingly important to make sure that they are correctly allocated to pharmaceuticals (as opposed to physician spending).

Jack Triplett described other examples of how spending by traditionally defined medical care industries combine into the price vector describing a specific treatment. Under present BLS procedures for cataracts—a case that was cited several times throughout the day—if the surgeries taking place in a hospital are sampled, then one set of price indexes would be generated for that; if surgeries shift to a clinic, then another set of price indexes would be obtained for that. If people switch from the more expensive hospital surgery to a less expensive clinic surgery, and if quality does not suffer, the ideal price index (from the perspective of the patient) would be capable of capturing the decrease in price.

Triplett continued, noting that one reason people have not thought much about substitution across medical care industries is that BEA data are organized at a higher level. BEA industry counts do not go down to the five-digit North American Industry Classification System (NAICS) level of aggregation, and therefore not all of the reallocation effects are readily visible.[2] He offered an analogy between the medical industries and the transportation equipment–producing industries to illustrate the industry-sector problem: the historical case of the automobile industry replacing the buggy industry. If the industry were defined as producing road transportation equipment instead of individual cars and buggies (although one might still want to get the prices of those), then in principle these substitution effects could be captured. If only the carriage and automobile industries were tracked, price and productivity measurement would capture only part of the effects that are of interest. When people found that it was cheaper per mile to go by car than by horse and carriage, and they switched from the latter to the former, the full price and productivity effects would not be completely explained by the indexes for either one. However, this problem arises when the interest is in welfare comparisons, rather than just in output comparisons. There is nothing wrong with the auto and buggy measures; rather, it is that aggregating them misses some of the welfare gains to the consumer. Triplett concluded that BEA would need to rework the way the five-digit industries are aggregated into the three-digit industries to do this—it is not only an index num-

---

[2]NAICS uses a six-digit coding system to identify particular industries and their placement in the hierarchical structure of the classification system. The first two digits of the code designate the sector, the third digit designates the subsector, the fourth digit designates the industry group, the fifth digit designates the NAICS industry, and the sixth digit designates the national industry (http://www.census.gov/eos/www/naics/).

ber problem. The usual way of modeling "substitution" in price index research does not adequately handle shifts of broadly defined products (like road transportation equipment or curing cataracts) between producing industries.

Triplett presumed that there are probably lots of instances of something similar happening, simply because a function is going out of one sector and into another one. Steven Landefeld agreed that these transitions probably do occur with some frequency. He noted that, wherever BEA has used quality adjustment methods, the focus has been almost exclusively been on final goods and services. That is, the agency has typically assigned the adjustment into the industry producing the final good. In this case, the goal is to examine how the change in the standard (expenditure) measure for gross domestic product (GDP) resulting from use of a new deflator works through on the industry (input) side.

Aizcorbe identified other characteristics of medical care that complicate the calculation of price indexes. For example, insurance plans vary in their payments for a given service, so patients in different plans effectively pay different unit prices. Ignoring bad debt and charity care, the uninsured probably pay the most for treatments and pharmaceuticals. When uninsured individuals turn 65, Medicare Part D comes into play and the drugs that they buy become cheaper. With aging, if the price that patients were paying before was high and the price that they pay once they join Medicare Part D is comparatively low, then the revenues that pharmacies or manufacturers receive for these drugs will fall. In this stylized case, nominal totals fall but the price indexes do not pick that up because they are tracking prices for, say, someone with Blue Cross/Blue Shield coverage and for someone enrolled in Medicare Part D separately. If this price index is used to deflate spending, a drop in quantity would be shown, even if the same population group—a portion of which has shifted from commercial insurance (or no insurance) to Medicare over time—is represented.

Aizcorbe suggested that handling this type of heterogeneity for deflation purposes needs to be analogous to the method for handling input substitution for the treatment of diseases. BEA would try to define the price as expenditures on all types of treatments by patients with all types of coverage and divide that by the number of patients. So in the population aging example, as people start spending less on drugs, it would be reflected as a price drop, not as a drop in quantity, which is exactly what is wanted for the national accounts.

In summary, the main reasons why BEA feels it needs to construct deflators differently from what is currently being provided by BLS in its Producer Price Index (PPI) program is that they want to be able to think in terms of treatment of a particular disease, not of a specific kind of treatment for the disease. Also, BEA would like the account to be capable of reflecting as a price change—and not as a quantity change—the different prices that patients under different plans pay for treatments as they shift from one plan to another. Finally, BEA would like to control for changing trends in the severity of conditions as well, to the extent that it is possible. The ideal would be to track over time a disease in which the severity of the condition is homogeneous.

3.2. BEA'S STRATEGY FOR COORDINATING THE INDUSTRY INPUT ACCOUNTS WITH THE DISEASE TREATMENT–BASED EXPENDITURE CONCEPT

Operating in parallel with the expenditure side of the NIPAs, on which the majority of the workshop discussion focused, is the issue about what to do on the industry side of the accounts. Once BEA begins deflating medical care spending by consumers using a new price index, the industry-side calculations must be revisited, as real spending on inputs in the production of medical care must equal real spending on final medical care goods and services. If the deflators on the spending side are wrong, it must be the case that the industry deflators are also wrong.

BEA's proposed approach to this issue is to reorganize its accounting structure for the medical care industry. The leading idea at the agency is to base this reorganization on a stylized model of health care in which patients are assumed to work through a care gatekeeper. Patients first go to their internist, pediatrician, or other primary care physician, who diagnoses problems and then sends them to different providers—the services are outsourced to labs, to professionals performing MRIs, to surgeons, and so on. Thought of in this way, the final good is provided by the primary care physician who orchestrates the medical care; everyone else in the system is simply an intermediate good.

Aizcorbe explained that, for national accounting purposes, this means that spending is deflated by the disease episode–based indexes that are allowed to cross NAICS industry lines, and the intermediate goods are deflated by PPIs. The gap between the real dollar amounts on the spending side and the amounts from the intermediate goods is attributed to the value added of primary caregivers. The critical distinction from the current accounting framework is that the specialists must be viewed as providing intermediate goods.

BEA's Plan to Revise the Medical Care Industry Accounts

In his presentation, Brain Moyer provided details about BEA's plan to modify the industry side of the national accounts so that they can remain synchronized with the disease-based organizational structure proposed for the satellite program. He began by explaining that, in addition to the accounts that register the contribution of consumer spending to real GDP, there is the less familiar set of accounts that show detailed inputs and outputs used in the production process by industry, of which health care is one. Here, real measures of value added by industry are established. Because health care accounts for a large and growing portion of the nation's economic activity—currently 16 percent of GDP—the importance of measuring its impact accurately and in a way that avoids major statistical discrepancies is self-evident. If a new measure of consumer spending is considered for the personal consumption expenditures (PCE), then that must be traceable back into the industry accounts to see which detailed components are contributing to that change.

Moyer began by explaining how BEA currently handles health care in the accounts (details of the BEA methodology can be found in Appendix C). BEA's input-output accounts show detailed transactions by industry, traditionally defined; so these transactions involve employees (i.e., consumers) who purchase health care services from various health care providers—physicians, hospitals, clinics, and so on. The health insurance industry is also viewed as providing a service to consumers (who often access the plans through employers). The GDP-by-industry accounts show real value added for the health care industries. In contrast to the proposed satellite structure, these industries produce and sell final outputs to consumers.

As noted above, the proposed change to BEA's framework will involve introducing a new primary caregiving industry. The primary caregiver industry purchases its inputs from the other industries providing health care—hospitals, clinics, laboratories, pharmacies, and so forth—which are viewed as intermediate purchases in the production of medical care. Other inputs, such as specialist physicians, could also be added. The idea is not new; its real-world counterpart is a health maintenance organization (HMO).

In his presentation, Moyer detailed how introduction of the primary caregiver category changes BEA's industry account picture. Employers still make contributions to employee health care plans, and employees still purchase health insurance; however, the new primary caregiving industry sells its output directly to consumers. This framework allows BEA to reconsider the definition of a unit of medical care and to incorporate disease-based price indexes. Most importantly it will allow the industry accounts and the NIPAs to be in balance, both on the nominal side as well as on the inflation adjusted or the real side.

Moyer presented a hypothetical example to illustrate how this additional industry reconciles the two sides of the accounts (Box 3.1). He suggested that the new accounting structure for the industry side is not only a mechanism to ensure that the accounts remain in balance but also has a realistic representational element. The real value added from the other providers is unchanged, which is to be expected, whereas there is an increase in the real value added for primary caregivers. This, he said, can be interpreted as resulting from the coordinating efforts that this newly defined primary caregiving industry is providing. In summary, as a result of moving to this new framework for the satellite account, the expectation is that there would be no impact on nominal consumer spending while the measure of real PCE and real GDP would increase. BEA has produced some initial estimates indicating that real consumer spending on medical care, measured in the new way, may be about 1.5 percentage points higher per year, and real GDP would increase by about 0.2 percentage points per year.

Discussion of the BEA Plan

During open discussion, Barbara Fraumeni—who, as a recent chief economist at BEA, has considerable experience with these issues—commended BEA for moving in parallel on both the expenditure and the production sides of the accounts, so that the inputs and outputs of the system would be identified. She summarized the key elements of BEA's plan, which she characterized as preserving the accounting structure while putting together an important disease data set that could provide the flexibility necessary to allow price indexes to potentially capture quality change effects. Fraumeni conceded that the current state of the art for measuring quality change at the disease treatment level does not do this quite yet, but she advised that this should be a high priority for the agency. Fraumeni suggested that BEA document its plans and progress through two papers, one that discusses the accounting system and proposed changes to it, and another that describes what can and cannot be done now and what the agency would like to tackle in the future—namely, the quality change issue.

Aizcorbe reminded workshop participants that BEA is still very much in the early stages of conceptualizing this industry-side structure and asked the experts assembled in the room to continue the debate on the topic. Extensive discussion followed about how the physician gatekeeper model would be operationalized in the national accounts. Fraumeni described the new model, which reroutes the way that expenditures flow through the national accounts, as essentially involving a "fake billing." The reason is that the billing does not come entirely from the primary caregiver, and it has no impact in nominal dollar value added or expenditures.

Sherry Glied agreed, noting that, even though integrated systems with gatekeepers exist in the real world (e.g., HMOs), medical care often takes place outside these systems. In medical care there are a few general contractors but, in practice, many people serve as their own contractor. From a mechanical standpoint, there needs to be a placeholder there, but it is not clear how it should work. In thinking about how to organize the industry-side inputs, Glied pointed out that sometimes there is a physical representative and sometimes it is a virtual idea. The products of the organizing industry are final episodes of treatment for specific diseases, but, she concluded, it will be a tricky task to figure out what belongs in that category.

Joseph Newhouse observed that insurance companies are already working to organize information at the disease level using the ETGs—even though their goals for doing so are somewhat different. They are beginning to actively analyze their businesses into what is called disease management, a new industry that is trying to manage chronic illness better. For the purpose of national accounting, Newhouse noted that the organizer or the gatekeeper is really a residual category to make the accounting entries balance. A mechanism is needed to capture the substitution that takes place, but then double counting has to be avoided, which is difficult when expenditures are fragmented then put back together again. Ralph Bradley added that, if the physician were truly an organizer, a grouper for the

> **BOX 3.1**
> **Hypothetical Reconciliation**
>
> On the spending side of the NIPAs, suppose nominal spending on medical care changes by $100. In the table below, the first column lists how the expenditure would be treated in the current framework; the next column represents the proposed framework. Currently, in the measures of personal consumption expenditures, the PPI—in this case, 1.07—is used to calculate real consumer spending. Since producer prices increased by 7 percent, the result is a $93 real change in consumer spending.
>
> **Treatment of a Hypothetical Expenditure, Conventional and Satellite Structures**
>
> | | Current | Proposed |
> |---|---|---|
> | Medical care: | | |
> | Change in nominal spending | $100 | $100 |
> | Price index | 1.07 | 1.05 |
> | Change in real spending | $93 = ($100/1.07) | $95 = ($100/1.05) |
> | Industry accounts: | | |
> | Change in nominal value added | $100 | $100 |
> | Primary caregiver | $20 | $20 |
> | Other providers | $80 | $80 |
> | Change in real value added | $93 | $95 |
> | Primary caregiver | $19 = $20/1.07 | $21 = ($100/1.05) –($80/1.07) |
> | Other providers | $74 = $80/1.07 | $74 = $80/1.07 |
>
> NOTE: Real value added computed through "double deflation."
>
> In the proposed framework, BEA would use the disease-based price index, which, in this example, increases by a slower growth rate relative to the PPI, which is consistent with some of the initial work done by BEA. This leads to a higher calculated growth rate in real consumer spending.
>
> To show how this flows through to the industry accounts in this example, the intermediate inputs of other providers are assumed to be zero. In the industry ac-

claims database system would not be needed, because physicians would be selecting the pathways of all the treatments and then reporting them. He also pointed out that, when reading a claims database now, there are claims for prescriptions and other medical services that are not assigned a diagnosis. This shows that the grouper fails to assign a medical purchase to a disease for each case.

The session concluded with discussion of Moyer's point that, if the deflator for medical care on the spending side of the accounts is changed, then the rate of GDP growth changes and a gap would open between that rate of real GDP growth

> counts, the goal is to derive a measure of real value added. To do that, the outputs and inputs of an industry must be deflated. In nominal terms, the value added is equal to the output of an industry minus its inputs. Output is deflated with an appropriate price index, as are the inputs, and real intermediate inputs are subtracted from real gross output, which provides the estimates of real value added. This is the typical way of computing real value added in a national accounting framework.
>
> Turning to the industry accounts side, under the current structure, the example shows a $100 change in nominal value added—nominal value added equals the change in health care GDP. For this example, $20 of that is attributed to the primary caregiver, $80 to other providers. The assumptions have not changed under the new structure; in nominal terms, everything adds up.
>
> Moving to the real side, under the existing structure, outputs and inputs are also deflated for the primary caregiver industry. Since a value of zero is assumed for intermediate inputs, the $20 for the industry is divided by the PPI, which gives a change in real value added of $19. Following the same procedure for the other providers produces a nominal value added of $80 and a real value added of $74. The sum totals $93 again, which matches the figure for real consumer spending on the other side of the account—the two sides balance.
>
> Moving to the satellite framework, characterized by the addition of the new primary caregiving industry, the other providers' value added is still $80; divided by the PPI, a real value added of $74 is obtained. Changing the structure does not change the value added of these other providers. The output of the primary caregiver industry is now the value added, or $20, plus the intermediate purchases it is making from the other providers, $80. Dividing this total by the new disease-based price index, hypothetically set here at 1.05, then subtracting the real value of intermediate inputs (the nominal, $80, divided by the relevant PPI, 1.07) gives $21. This total, plus $74, equals $95—the two sides of the account are in balance.
>
> SOURCE: Workshop presentation by Brian Moyer.

and the measure of change in the industry accounts (if left unchanged). This gap has to be entered as a line item somewhere; the issue is then how to interpret it. Landefeld reminded participants that this idea of making an adjustment in a top-level category on the industry side of the account, to make it balance with the expenditure side, is not unprecedented. For computers, this is how any gains in productivity are categorized. The residual is attributed to the top-level industry—computer manufacturers, not the component manufacturers.

Triplett also framed the industry side accounting issue in the context of productivity, which, like the national accounts, offers a logical construct for organizing data. If indeed the primary care physician industry is to be where all excess multifactor productivity[3] will be categorized, he urged BEA to be explicit about the method for treating the residual difference between the expenditure-side and the industry accounts under the old and the new frameworks. This level of detail underlies the accounts, but the numbers published by BEA are for the health care industry, which is the aggregation of the subindustries—the hospitals, the doctors' offices, and the rest—for which data are supplied by the Census Bureau and BLS. Triplett also suggested that BEA will have the problem that price indexes for the subindustries will not equal those for the aggregate level—there will need to be a reallocation term. Based on the Moyer presentation, BEA now has a story to tell about the source of economic activity driving the inequality, which, in some other contexts, is called a statistical discrepancy.

In discussions of the options about how the inequality could be handled in the satellite account (other than the gatekeeper productivity mechanism), David Cutler made the point that one possibility would be to simply create a line in the accounts called "total factor productivity" and not specify which industry gets it. In addition, it is sometimes important to know how productivity is affected in overlapping disease areas below the medical care industry level. As an example, if the medical profession gets better at treating heart attacks, it might suggest that diabetes treatment has improved, since one of the things these patients die of is heart disease. Cutler observed that in one sense this is right, but in another it is not. It is right in the sense of the broad medical care industry, because people with diabetes are now living longer and their quality of life is better. In other ways, it is wrong, because it is not the treatment specifically for diabetes that has led to the improvements. So, for various purposes, productivity gains need to estimated at different levels of aggregation.

The key is to make sure that the pieces add up to the total; otherwise, the person who receives treatment for heart disease may be double counted in the estimated productivity improvement from that, as well as from treatments of diabetes, high cholesterol, and hypertension. Only with an accounting structure can one ensure that entries do not appear in multiple places or, if they do, that they are parsed to add up correctly. This is done by looking not only at the expenditures on and productivity of the treatment of the disease, but also at the productivity of each particular input; so it is not just the productivity of treating the heart attack, but also of the hospital and of the physician and of the pharmaceutical company.

In response to Cutler, Landefeld made the point that, to allocate the adjustment across industries or subindustries begins to call for a lot more work and, of course, all that has to be transparent. He noted the similarity here with the discussion that has gone on for years about the statistical discrepancy in the accounts.

---

[3]Multifactor productivity is defined as output generated per unit of combined inputs; inputs may include capital, labor, energy, materials, purchased business services, and so forth.

PRICE INDEXES   41

Many users would prefer that the accounts categorize the residual in one place, because once it is allocated, people want to know explicitly where it went and how that flows through to multifactor productivity. He continued, saying there may be some value in just leaving the residual in one place until it can be figured out where it properly belongs. In other words, final expenditures is what is in PCE, and that needs to be made clear. With the Moyer presentation, BEA laid out what that implies for the industry accounts.

3.3. TRACKING QUALITY CHANGE OF MEDICAL GOODS AND SERVICES

In a world of ideal measurement, BEA's satellite health care account would be deflated using quality-adjusted price indexes, as is already done for the most methodologically evolved components of the NIPAs. This is the logical final step in the program, but work is currently too embryonic for BEA to go that far now—things are very much in the research stage.

Matthew Shapiro laid out the (long-term) agenda and identified hurdles that would be met along the way. He began by noting that, in constructing a constant quality price index intended for use in measuring real change in economic activity, the idea is to hold the mix and character of goods and services constant. For the health sector, price inflation is important, but there are also huge changes in how medical services are performed. The only constant is that the treatments are intended to improve the health of people with a given condition.

Shapiro drew from the example of a treatment he has worked on, cataract surgery, to illustrate the sometimes subtle components of a price change. If a doctor or a facility charges more for a specific treatment, that should clearly be counted as a price change—that is the easiest case. But, he continued, the price change that accompanies a facility shift should also sometimes count. If a procedure previously performed in a hospital is now done in an ambulatory care facility or an office and the price is different, the price index should capture that.

Here, Shapiro noted, is where some differences in method emerge across the agencies. BEA's preferred approach is treatment oriented; whether a patient has a cataract procedure done at a hospital or at an ambulatory care facility, if the outcome is the same, any observed price change should be treated as such. The BLS indexing procedures have typically looked only within each type of facility, and the aggregation is done in a way that does not capture this kind of price change. Similar situations arise elsewhere. Consider cases in which a health plan reimburses only a fraction of the price change, or for which Medicare decides to change the amount that it pays doctors. Shapiro argued that, as long as the doctor is performing the same service, the change in what gets reimbursed should be reflected as a price reduction.

In addition, situations arise in which the way a procedure is actually done changes. For example, staying with the cataract case, modern treatment has gone from a sutured to a sutureless procedure—there has been a technological change in how the surgery is performed. The fact that patients do not have to pay for the suture under the new procedure should be considered a cost reduction, assuming that it is just as effective. In the language of price index construction, this is a debate over what gets linked and what does not. BEA is hoping to develop a method in which such factors as changes in the mix of inputs, in who pays, and in the location of the service are not relevant to pricing the unit of measurement for cases in which these things have not affected quality. Shapiro views this as correct if the goal is to measure real prices for purposes of deflation of the national accounts.

Next, Shapiro asked how this type of price index construction might be conducted on a larger scale than has been done to date in studies by health economists, which have typically focused on treatment of one specific disease. He suggested that it will inevitably require compiling data from bills on costs of episodes in a detailed, broad-based, and systematic way. The Consumer Price Index program for measuring health costs already involves sampling bills and repricing them from period to period. For the cataract surgery case, BLS would draw a random bill and note the portion going to the doctor, to the facility charge, to materials, to nursing, to anesthesia, and so on. Price change is measured asking the facility to reprice that randomly selected bill in each period for a hypothetical patient (with the same payer) who had the same mix of treatments, applying recently observed component prices. Over time, new bills are periodically drawn that may reflect changed processes or input mixes, but this only has an effect on the index going forward. In the cataract example, the fact that the suture disappeared may get missed in the price system.

The BLS-developed method of pricing bills for specific, well-defined kinds of treatments marked a significant improvement over indexing methodologies that simply priced care components, such as an hour of a doctor's time or a day in a hospital. Shapiro pointed out that, in fact, what BLS has done for the past decade or so sounds a lot like what BEA is proposing with its large medical care database. However, the BEA idea is more ambitious, in that it would attempt to account for the fact that the suture in the cataract has disappeared, or that the service has moved from a higher cost category of provider—a hospital—to a lower cost one, such as an ambulatory care facility. Shapiro embraced the BEA proposal, which he described as fundamentally following the treatment of diseases, not a set bundle of inputs. He added that moving to this type of framework—really developed by Newhouse, Cutler, and others on a case by case (or disease by disease) basis—en masse in a systematic way would represent a major move in the right direction.

BLS participants also described their agency's progress on this kind of price measurement. Bonnie Murphy noted that the PPI program can handle substitution within providers. So, if a cataract surgery was changed from a sutured to a sutureless procedure, and it was performed by the same kind of provider—say, in the hospital—that substitution could be captured. If the nonsuture cataract surgery was performed in a physician's office, that would not, as BLS current index sampling procedures cannot accommodate substitution across providers.

In the examples cited at the workshop, the advantage of moving to a structure organized by major disease categories rather than by provider became clear. Murphy agreed, stating that, theoretically, BLS would want to be able to measure price change associated with these kinds of substitutions. She added that, in their experimental indexing work, described in detail in Section 3.4., they plan to allow substitution across provider types. Furthermore, if suitable outcomes information becomes available, BLS would look into methods to do a quality adjustment; without that, an effort would be made to do a direct price comparison, which would show the price change taking place between the two periods.

Shapiro recommended that BLS document their work, both to show the extent to which the current CPI and PPI programs are equipped to handle input and provider type substitutions, as well as the plans for alternative indexes designed to move further toward a disease-based unit of measurement. He noted that the past CPI literature includes articles that have helped spur important changes in price indexing methodology and practice. He added that both the CPI and PPI should produce papers prior to the workshop, his impression had been (perhaps mistakenly) that not much was happening on this front.

Shapiro concluded that, at some point, it will be time to move on to the quality change issue; following the cataract example further, it is important to ask whether losing the suture was a good or a bad development. In reality, it probably improved treatment outcomes by reducing the likelihood of mistakes and complications from having to put in a suture. Treating complications adds to cost, which should be picked up in the accounts. That these costs are eliminated should count as a productivity improvement, which would have to enter the accounts through an adjustment to the price index. However, there may be instances for which losing something from a bill might not be a positive in terms of patient outcomes—disappearing inputs do not necessarily always constitute a price reduction. This is precisely why medical experts are needed to evaluate the constant quality caveat, even if no explicit quality adjustment is taking place. BEA has, for the moment, put quality adjustment aside, which Shapiro said was reasonable. Presumably, down the road, indicators of mortality and morbidity will become increasingly available and, perhaps, progress can be made to fold this information into quality measures on a case by case basis. But those, Shapiro agreed, are two steps that could be worked out in sequence, as BEA has proposed.

In her comments, Barbara Fraumeni suggested that BEA also ramp up documentation of plans for its satellite health care account, specifically about possible quality adjustment to their price and quantity estimates, and a defense of their view that the task is separable from the disease-based expenditure allocation work. She worried that, if BEA defers this effort, they may find out after spending a huge amount of time working with these very large databases that, had certain fields (e.g., perhaps those related to product quality review) been saved, it would have been much easier to come back and do some sort of quality adjustment.

Fraumeni said that BEA must recognize that quality adjustment is important and that one way to do it is through outcomes assessment. She advised BEA to take the "house-to-house combat" approach with respect to quality adjustments; there may be some diseases for which it is fairly well known that there have been significant changes in the types of treatments and the efficacy of those treatments—mental health might be one of those. Work could gradually proceed on a case by case basis to begin making gradual adjustments in prices and quantities of these treatments in a way comparable to what happened when BEA began dealing with quality adjustment for computers.

On behalf of BEA, Aizcorbe welcomed these suggestions. She was particularly interested in getting a sense of the extent to which the staged strategy—first tackling the quantities and expenditures, then quality adjustment—would fare in the end. She assured Fraumeni and others that the agency was keenly aware of the quality adjustment issue but that, at the moment, they simply do not have a systematic way of dealing with it. Fraumeni responded that perhaps there were some cases on which BEA could get started sooner rather than later, but that they would need to be selective.

In summarizing what he heard during the session, Landefeld conveyed the view that work in progress at BLS would ultimately help capture differences and changes in the quality of care relative to some best practice across regions or types of hospitals, but that they would tend to miss changes in best practices that take place over time. He said that how to begin measuring quality improvements, whether they occur in large discontinuous jumps or incrementally, was the key question for which his agency would need help from the assembled expertise at the workshop. Landefeld then asked about the effect of including quality adjustment, going from a conventional price index to the proposed version needed for the satellite account. His view was that BEA could in fact move significantly toward where it wants to be by getting the cost (nominal expenditure) component correct first, and then appealing to external guidance for suggestions on how to go forward on quality adjustment—not cross-sectionally, but with respect to those large discontinuous changes in technology.

3.4. THE ROLE OF THE BLS PRICE INDEXES

Michael Horrigan, associate commissioner in the Office of Prices and Living Conditions at BLS, provided introductory comments for presenters from his agency's CPI and PPI programs. The presentations focused, in part, on the kinds of price changes described by Shapiro and the extent to which they could be captured by BLS field procedures in a timely fashion. In introducing his colleagues, Horrigan said the purpose of the presentations was to provide a sense of the agency's plans; he acknowledged that whether or not the PPIs or CPIs would meet BEA's needs to proceed with its satellite account was an open question.

One important consideration is the frequency of the reporting requirement to the public, which has an impact on the range of methodological options. For much of its work, BLS has to think in the context of a monthly production process. Horrigan pointed out that the Cutler-Rosen work measuring per-person costs of completed episodes can take advantage of less frequent reporting requirements. The varying timeliness constraints may lead to different decisions regarding, for example, whether an episode-based approach or an encounter-based approach is appropriate. Likewise, some of BEA's objectives with its satellite account allow for less timely periodic analysis. The fact that some version of the national health accounts could be issued on an annual or quarterly basis creates extra possibilities that should be exploited.

CPI Medical Care Price Indexes

The workshop sessions covering initiatives to advance medical care price indexes informed questions about how and in what ways BEA will be able to draw from BLS sources to deflate nominal expenditures for estimating real GDP for the sector. Ralph Bradley from the CPI program opened the discussion presenting preliminary results produced by experimental work on a medical care index organized by disease. He stated that the motivation behind the initiative was to improve the medical care CPI by trying to capture protocol changes, such as those exemplified by the cataract or mental disorder cases discussed by workshop participants. Bradley also emphasized the importance of determining the feasibility of real-time production and publication of disease-based price indexes.

Bradley began with a description of data requirements. He stated that BLS would not be in a position to initiate a new survey in the near future, so research investigating the feasibility of disease-based indexes must rely on existing data sources. MEPS, because it captures medical expenditures and also measures medical utilization, is the obvious candidate to use in the CPI. Its level of detail facilitates price index work that encompasses many of the needs; for example, substitutions toward less costly inputs for a given treatment should be reflected in the data. MEPS is analogous to the Consumer Expenditure Survey used to generate weights for many CPI components in aggregating to the all-items index. Annual data from MEPS would be used in much the same way to generate weights for the disease-based indexes.

Bradley presented preliminary results emerging from initial analysis of the MEPS data. These results are summarized in the tables in Appendix A. For this research, Bradley and colleagues merged the MEPS conditions file, which lists each diagnosis an individual receives, or the diseases reported on the household file, with the MEPS event file. The events file includes various treatments received, office visits, emergency room visits, outpatients encounters, pharmaceuticals, etc. The MEPS data are annual, so a monthly inflation number had to be created. For this, weighted CPI data for existing categories—office visits, hospitals, pharmaceuticals, etc.—were used to generate monthly price growth estimates. The weights in MEPS are updated annually for all inputs used in each disease treatment. If substitution away from a more expensive input to a less expensive input occurs, the updated weight will reflect that.

Substitution among medical protocols is one aspect of the analysis. Another, reported Bradley, has to do with the fact that the population is getting older and, with that, many disease treatments are increasing in intensity. Table 3.1 shows trends in average utilization per person (not per disease) for a set of services, as reflected by data from the MEPS consolidated household file. Each year, the representative individual is getting a little older, by roughly 0.2 year. Although there is little trend in hospital utilization, there is an increase in utilization for the other areas, particularly for pharmaceutical. The rapid growth in emergency room utilization could be, in part, a function of the increased fraction of sampled individuals who are uninsured and who therefore rely more heavily on that option.

Bradley next reviewed methods for handling comorbidities. The research group looked first at the mean number of diseases treated per office visit, linking the conditions file with the events information files on office visits. They found that the mean number of diseases treated per visit increased from about 1.5 in 1996 to over 2 for 2004. One approach considered by the group for treating these additional comorbidities involved calculating a proportional distribution. The idea here is that if, in one year, a patient with diabetes has a physician visit, that entire encounter is allocated to diabetes. If, in the next year, the patient now has diabetes and a heart condition, that will be prorated—half of the visit will be allocated to diabetes and the other half to heart disease. If a prorating scheme is used across the treated diseases, increasing comorbidities will increase service productivity. Doing no prorating, however, may bias the index upward, because a service is double counted. So far, BLS has generated only indexes without pro rating, but they plan to produce additional indexes with prorating.

Bradley also described item weighting procedures. For purposes of the CPI

TABLE 3.1 Average Service Utilizations, 1998-2004

Year	Hospital Nights	Inpatient Discharges	Physician Visits	Outpatient Visits	Emergency Room Visits	Number of Filled Prescriptions
1998	0.582	0.098	4.569	0.461	0.160	7.202
1999	0.513	0.100	4.412	0.428	0.157	7.480
2000	0.592	0.102	4.390	0.470	0.169	7.783
2001	0.572	0.106	4.765	0.526	0.194	8.773
2002	0.562	0.102	5.110	0.558	0.196	9.345
2003	0.557	0.101	5.144	0.544	0.192	9.640
2004	0.567	0.103	5.219	0.513	0.193	10.003

SOURCE: Workshop presentation by Ralph Bradley.

program, the methodology is driven by the intent to price the goods and services financed by the out-of-pocket payments of consumers. Relative to, say, total system expenditures, this produces a different set of weights on hospital services, physician services, and outpatient services. If a shift occurs from hospital to outpatient services, there may be very little savings in terms of consumer out-of-pocket expenditures; most of it may accrue to third-party payers. This will of course affect the relative performance of the different kinds of indexes.

For the experimental medical CPI, BLS organizes disease treatments into the following categories:

1 - Infectious and parasitic diseases
2 - Neoplasms
3 - Endocrine, nutritional, and metabolic diseases and immunity disorders
4 - Diseases of the blood and blood-forming organs
5 - Mental disorders
6 - Diseases of the nervous system and sense organs
7 - Diseases of the circulatory system
8 - Diseases of the respiratory system
9 - Diseases of the digestive system
10 - Diseases of the genitourinary system
11 - Complications of pregnancy, childbirth
12 - Diseases of the skin and subcutaneous tissue
13 - Diseases of the musculoskeletal system and connective tissue
14 - Congenital anomalies
15 - Certain conditions originating in the perinatal period
16 - Injury and poisoning
17 - Other conditions
18 - Preventive services without diagnosis

Preliminary results for all of the total expenditure and out-of-pocket expenditure disease-based indexes are presented in Appendix Table A.5. Bradley illustrated how the index works using mental disorders as an example. For this kind of service, the mean number of office and outpatient department visits has dropped dramatically over the study period, but the mean number of prescriptions has increased. Using the current methodology, in which each individual service is tracked separately, these shifts go unaccounted for and a growth rate of 37 percent emerges for the price index for treatment of mental disorders over the 1999-2004 period. When the annual quantity updates are performed, the price index growth declines to 7 percent. Bradley noted that mental disorders is an extreme example—a case in which the substitution effect is pronounced. It does provide an example, though, of the kind of impact that can occur when changes in protocol are folded into the broader based indexing method.

Overall for medical care, when a total expenditure concept is used, a price index that uses annually updated protocol changes grows less rapidly—by about 3 percent—than does an index that uses only fixed quantities. But, for the out-of-pocket scope index, the reverse happens; the fixed quantities index grows less rapidly than does the quantity updated index. There are several reasons for this. Sources of financing play a very important role. As Appendix Table A.4 indicates, the out-of-pocket share of inpatient hospital expenses is comparatively small. Bradley reported that it is common to see situations in which going from inpatient to outpatient status produces a 50 percent drop in the total cost of a medical procedure. But, simultaneously, patient copays may rise, and there could actually be an increase in terms of out-of-pocket payments going from inpatient to outpatient. Also, hospital services is the stratum for which prices are rising most rapidly. So, if a total expenditure approach is used, more weight is being put on the hospital index relative to an out-of-pocket index.

Next, Bradley described the BLS plans for continued research, specifically to develop treatment-based indexes that use a CPI scope. The CPI includes out-of-pocket payments made by individuals or families (including premiums for employer-provided health insurance paid by employees), plus the insurance payment financed from the employee's Medicare Part B payment. Jack Triplett commented that, although it is a little hard to conceptualize exactly what a price index for out-of-pocket expenses will look like, it is the relevant concept for the CPI. He made the point that it is odd that, in the past, the CPI program has spent time trying to figure out how to price the cost of insurance when the CPI weights reflect only the out-of-pocket expense portion. The out-of-pocket scope index is a little more difficult estimate, relative to the total expenditure scope index, because a crosswalk has to be created between the Consumer Expenditure Survey data and MEPS data, which include the third-party payments.

Finally, BLS will looking for ways to better differentiate medical services in its indexes. For example, currently, the unit of measurement for office visits is the visit itself; however, it should be possible to take advantage of data fields in MEPS indicating in more detail the types of services that are provided, such as an MRI, an X-ray or other things, and this may lead to better measurement of quantity and possibly quality.

PPI Research Plans for Medical Care Price Indexes

Bonnie Murphy presented the BLS plans to develop a "multi-industry price index structured by disease." At the moment, these plans are very much still in the research phase. She emphasized that the PPI program will not change the way it measures prices—it will not be pricing an entire disease treatment episode. Since the alternative structure uses PPI data that are already collected (there will be no new surveys), it is a low-cost experiment. This first new product is scheduled to appear in 2010 at the earliest, mainly because of restrictions on the data that BLS can get to implement this project. The PPI program also has plans for a second project to push forward on ways to quality adjust for hospitals. This research project is in the public comment phase right now; it is possible that, if the feedback on the concept is positive, a new method of quality adjustment for the hospital index could be implemented very soon.

The medical price indexes in the PPI have to be flexible to meet a range of user needs; for example, the industry or provider-based indexes have been needed by BEA for industry-based deflators. Indexes organized according to payer meet the needs of health insurance companies and other public and users (such as CMS). Also, BLS receives many requests from health insurance companies asking about price trends for Medicare and Medicaid; timely (monthly) and comprehensive price indexes generally meet these needs. In part because of these clients, the current industry indexes will not change. Workshop participants referenced data sets that they have used in accounting exercises that are available only with a 2- to 4-year lag; with these, it would be very difficult to come up with a product that is very timely and comprehensive. In addition, Murphy stated that BLS would not publish anything that did not cover the entire population.

Murphy began by acknowledging that a major motivation for working on the disease-based indexes is to help meet a research or national health account need. This alternative index she hoped would fill a gap by providing timely price data on a path or course of treatment for any given diagnosis across all providers (industries). Ideally, it would be capable of capturing substitutions of treatment protocols within and across treatment providers—for example, the cataract surgery that has changed from inpatient to outpatient (discussed in Section 3.1.), or the ulcer treatment that has changed from hospital treatment to drug treatment.

Murphy then described the current structure of medical service PPI components. Currently, PPI publishes indexes for the set of medical care industries shown in Table 3.2.

Murphy reported that the PPI for hospitals has a relatively robust sample, which can support publication of data for Medicare and Medicaid patients, as well as 23 additional indexes by major diagnostic category for non-Medicare and non-Medicaid patients, so BLS already publishes this part of its health care indexes by disease. The pricing unit itself consists of expenditures on the specific procedure. For example, for the patient who had an appendectomy surgery in the hospital, the price would be the reimbursement that the hospital receives from any payer, whether Medicare or Medicaid, private insurance, the patient, or any combination of these payers. The PPI captures the total reimbursement (from admission to discharge) for this appendectomy. Specifically, the PPI captures the reimbursement for all of the services that are included on the bill. If the physician who performed the surgery bills the patient separately, then those revenues would not be a part of the hospital bill and would be given a chance of selection in the physician industry.

It is important to clarify that, if the patient goes to a physician's office for

TABLE 3.2 Publication Dates for Medical Service PPI Components

Industry	Publication Start Date
Pharmaceutical preparation manufacturing	July 1981
General hospitals	Jan 1993
Psychiatric and substance abuse hospitals	Jan 1993
Other specialty hospitals	Jan 1993
Offices of physicians	Jan 1994
Diagnostic imaging centers	July 1994
Medical laboratories	July 1994
Nursing care facilities	Jan 1995
Home health care	Jan 1997
Retail pharmacies and drug stores	July 2000
Health and medical insurance carriers	Jan 2003
Residential mental retardation facilities	Jan 2004
Blood and organ banks	Jan 2007

SOURCE: Workshop presentation by Bonnie Murphy.

diagnosis of the appendicitis, that office visit would be included in the PPI physician index, which is separate from the hospital index that captures the price of the surgery (recall the example above for how the PPI handles reimbursements). The PPI segregates by provider, however, physicians who operate out of hospitals, HMO medical centers, or similar facilities, and bill separately are included in the physician PPI. The immediate plans for the PPI, for July 2008, call for the general hospitals index to be published according to major diagnosis category (MDC) without payer detail. PPI will also publish an alternative index by payer type only—that is, by Medicare, Medicaid, and "all other payers," again without MDC detail.

Future plans call for aggregating the indexes by the Census Bureau industry weights, which is what is typically done in the PPI. A much more dramatic improvement, which may be made possible by the 2007 Census Bureau implementation of the North American Product Classification System structure, may allow the PPI (by around 2010) to publish alternative indexes that cross seven health care industries: pharmaceutical manufacturing; all hospitals (general, psychiatric, and specialty); physicians; medical laboratories; and diagnostic imaging centers.

This plan would allow the indexes to capture some cross-industry medical service input substitution. Murphy provided an example to illustrate how the new price index would work. She returned to the example of an eye surgery that moves from an in-hospital setting to a physician's office, and that is now performed at a lower cost. The new (in office) treatment for this eye disease enters the market in year two, and it does not initially represent a large portion of the market. The current PPI would not show price change in the physicians index when it enters the market; the same is true for the hospital index. The old treatment is still occurring in hospitals, so PPI would still price it, and it will continue to be included in the hospital indexes for a number of years.

Under the alternative indexing methodology, both treatments (the traditional hospital-based one and the new office-based one) would be concurrently priced until some threshold is reached, defined by a specific level of market penetration for the new treatment. In the alternative index, the hospital treatment would be eliminated altogether from the sampling at some point; BLS is currently in the research phase of determining exactly when this would be—and what the threshold would be. The index would show the price change between the hospital treatment and the physician treatment at the threshold time period.

Although Murphy expressed optimism in the proposed design for the experimental index, she was also careful to list limitations to this alternative structure:

- The PPI covers only 13 health care industries; there is no coverage, for example, of ambulance services or of outpatient surgical centers, which tend to be important in substitutions between hospitals and physicians.
- Disease-based structures based on the North American Product Classification System are unavailable for some providers (e.g., home health care). The PPI will be limited as to the number of industries it can include in the alternative structure, so something will have to be done about expenditures that don't fall into these categories.
- The PPI will publish only disease categories with sufficient item data coverage. The PPI cannot show price change for items in the alternative price index at least until additional economic census data become available.

The experimental PPI would still assume that the outcomes were the same before and after the change in protocol—for example, from hospital outpatient to physician outpatient. Of course, this will not always be the case, and ideally one would want some quality or outcome assessment. Murphy stated that a high priority is to get a quality adjustment assessment in the index as soon as possible.

At the moment, the PPI does not have a systematic method for quality adjustment, although Murphy was optimistic that, at some point, the CMS Hospital Compare data set could possibly be used to quality adjust the hospital index. There are a handful of conditions included in the Hospital Compare data set—for example, heart failure and shock, pneumonia—for which data are collected from hospitals on a quarterly basis. The program has established some quality indicators developed by clinicians that are likely to be better than anything that could be developed by BLS which does not house clinical expertise. The underlying methodology needs more research, but if it is determined to provide an acceptable basis, the capability to quality adjust will grow along with the Hospital Compare data set. The current PPI plan is to quality adjust all treatments, on an annual basis, for which data are collected at CMS. Murphy cautioned that this was a small step in quality adjustment—for example, it applies only to hospitals. But because the CMS data are longitudinal and clinician approved, they are worth considering for the program.

Open Discussion of BLS Plans

Triplett commented that, in 1992, the PPI released its then-new hospital price indexes, which were a great advance over what had been done historically. As noted above, for many years, the unit of measurement for BLS price indexes were things like the cost of a day in a hospital. The PPI's advance was to move toward the episodes-of-treatment concept, in which a diagnosis for in-hospital treatments would be priced out initially and then followed. It is a synthetic price that is estimated by asking the hospital what it charged for a diagnosis that has the same characteristics, the same demographics, and other conditions. The improvement resulted in a price index that grew less rapidly than the older index.

Now appears to be time for the next major improvements to BLS price index programs. Triplett summarized three dimensions to the upgrade, noting the encouraging development that BLS is proposing work along all these lines. First is the need to adjust for improved (or deteriorated) medical treatments. Everybody, including BLS, agreed that it would be much better if quality adjustments were made to reflect these improved treatments. The second upgrade to the system is to extend the general approach of pricing episodes of diseases to nonhospital indexes. The third is to follow and perhaps adjust indexes when a treatment moves across facilities—or industries, as the structure is set up now. Currently, the hospital is an industry, the doctor's office is an industry, and clinics are an industry.

Next, Triplett spoke in more detail about BLS quality adjustment plans. The agency's proposal follows the usual PPI method for making quality adjustments. The PPI is a constant input, fixed technology price index, just as the cost-of-living index is a constant utility, fixed preference function index. The CPI uses consumer preferences as a way to value a change in medical treatments, but the PPI theory on this is based on production costs. So the theoretical ideal is to use the difference in production costs between the old treatment and the new treatment to make a quality adjustment in the index. Triplett expressed skepticism about these PPI plans working because this theory of the output index embodies a problematic conceptual approach.

Because the PPI is in theory a fixed input, fixed technology index, it is consistent if the quality change does not involve a change in the underlying production technology; Triplett pointed out that a lot of quality changes fit this model. For example, computers have over some periods used the same technologies, but there have been improvements that make machines faster. In the medical care context, many of the trends of interest involve new technologies. The cataract surgery that moved to a sutureless procedure is a good example. It is an innovation, not a constant technology process. This innovation not only improved the treatment, but actually reduced its cost.

One could take the old technology and ask what would it cost to produce the characteristics of the new treatment in the old technology. The problem, Triplett continued, is that the outcome could not be produced using the old technology. On this topic, Triplett concluded, the procedure described by Murphy might work well for some of the limited purposes for which it has been proposed, but it will not get at the major changes that concern most people working on productivity change in medical care. Those are the big changes in medical technology for which there is not a consistent cost estimate for the new and the old technologies. Again, Triplett praised the presenters for trying to do something about this, but cautioned that there were these limitations in terms of the ability of the methods to pick up the new technology-driven quality changes.

Next, Triplett commented on differences between the grouper-based and PPI approaches for handling major changes in medical treatment that involve new technologies. The BLS method of handling the case in which outcomes before and after the change were not equal would involve a linking procedure. The alternatives discussed during the workshop are unit value indexes, in which a direct comparison is implicitly made between two different treatments, and indexes using linking methods, wherein prices of the two treatments are linked and not compared directly.

In the case of the direct comparisons, changes to the good or service are ignored. A generic drug is treated as equal to its branded equivalent, and its price is compared directly (the current CPI procedure). The measured price drop is too large for the typical case in which some people do not switch to the generic version. The error arising from the direct comparison depends on the magnitude and direction of the quality change since all of it has been incorrectly been called a price change.

The linking method is more complicated because the price change is implicitly assigned from the things that changed (it is not true that the linking method implies the exact opposite of the direct comparison method—that is, that all quality change is ignored). The error occurs, roughly speaking, when prices are rising or falling, and some of the price change is attributed to the product quality change. So, Triplett raised the question, how are we to know what is the right way to do it? Comparing the generic and the branded drug directly might be better than ignoring the price change that occurred when the generic was introduced, even if direct comparison contained an error; but, we should strive for better methods. He suggested that the right way is to avoid constraining oneself to either making a direct comparison (with the implication that there is no quality difference) or to using a linking technique (implying little price change).

Triplett argued that an explicit adjustment is the preferable method for adjusting price indexes to account for changing treatment quality. The explicit quality adjustment in the case of medical care requires information on medical outcomes, which throws the problem back to the fact that little is known about them. A small number of studies are cited over and over because that is all that has been explored; and, even in those studies, none used explicit outcomes measures. One can only conclude that a lot more work will have to be done before this kind of approach can be implemented in a broad based way throughout a statistical program. Yet, Triplett concluded, it is the right course for future work.

Triplett's final point was about the issue of following a treatment across industries, providers, or facilities. This could be done under the BLS method of collecting data from providers. If outcome measures were available to allow quality adjustment, that could also be done across provider classes. Returning to the cataract example, if BLS knew that outcomes from surgery in the outpatient and inpatient treatments were the same, they could make this direct comparison. But there are so many cases for which this is not known. Without the research, one cannot be sure that reducing the number of days after a normal birth delivery yields an equivalent outcome or whether it is just an attempt to reduce costs. This implies a major research agenda; figuring out how to do the quality adjustment requires a lot of scientific and medical information.

The Illustrative Case of Pharmaceuticals

Patricia Danzon commented on the BLS presentations, drawing insights from one of her areas of expertise. In the case of pharmaceuticals—as with standard cross-national comparisons that have been done for other services, such as hospital days and physician visits—the practice has been to simply divide expenditures by number of units to infer difference in prices. The expenditures are hugely different, the quantities are similar, and the inference has been that it is therefore the prices that are much higher in the United States. It is these misleading results, Danzon stated, that make pharmaceuticals a good example of why accounting for quality differences is important in medical care price indexes.

The issue, then, is to determine how much of the price difference observed in these statistics is really the services and the quality. Because pharmaceuticals are precisely defined—they are measured at the level of the mechanism of action, the strength, the pack, the manufacturer, etc.—Danzon has been able to calculate accurate comparisons of utilization and price differences across countries. She and her colleagues have found that a significant portion of the expenditure difference across countries is explained by variation in the drugs being used—the formulations that may have quality dimensions to them. Clearly, simply dividing expenditures by number of prescriptions can vastly overstate price differences.

Danzon raised the issue of international comparison as it relates to the discussions by Triplett and Shapiro. For pharmaceuticals, the tendency to use number of prescriptions is essentially imputing all the expenditure change to a price change (the direct comparison), whereas in reality much of it is in fact attributable to new drugs or new formulations and of course the generics. If a CPI tracks volume purely based on the number of prescriptions, it will be upward biased when there is change in technology that is quality improving. She noted that, to her knowledge, BLS was handling generics appropriately now, in that they are being treated as equivalent to the branded products that preceded them.

An important gray area is the changes in drug formulation that occur around patent expiration that involve strategies by the branded manufacturers to extend their patent life; for example, the firms may delay the release of new formulations and, instead, introduce single isomer versions of the original drug, or combination products. In BLS's current procedure for drugs, Danzon suspects, these new formulations or combinations are treated as new products, which may not always be appropriate and may lead to a bias in the price indices.

A common manufacturer strategy is to raise the price on the old formulation that is going off patent relative to the new formulation, in order to encourage people to switch to the newer products. Thus, if a price index is being used that tracks the older standard dosage form but does not pick up the delayed-release version that is in fact becoming the norm in the market, it will overestimate the rate of price inflation, since it includes the formulation that is no longer being used very much in the market. When the market baskets are updated, this will be picked up but, in some cases, the delays are significant. Danzon reported that, in these cases, her own research indeed found a much more rapid price increase for the formulations when they were going off patent and being replaced by versions that still had some exclusivity.

Danzon also raised the issue of pharmaceutical invoice rebates. When prices are sampled at the hospital level or at the pharmacy level—or, for the PPI, at the manufacturer level—electronic rebates that go directly to payers will be missed for the outpatient pharmaceuticals. These rebates reduce the price to consumers and, in turn, the revenue that the manufacturers get. This is important for the branded products (for the generics, manufacturer revenues will be correct). What may be misreported is the amount that consumers pay, because the discounts go to pharmacies.

How much that rebate to the pharmacies then gets passed on to consumers in the form of price reductions of other goods and services is unknown. She commented that, it is mostly an article of faith among economists that much of the rebate that comes from the manufacturer to pharmacy benefit managers is passed back as part of the cost of the drug benefit to the employer and therefore ultimately to the consumer. The evidence that she has seen (a study by the Congressional Budget Office) is that something like 80 percent of these drug-specific rebates were passed back to the employers. If they are not passed back, it would be picked up more in the cost of health insurance—it will not show up in the pharmaceutical component. Danzon noted that these rebates are not trivial amounts. For the generics, the average rebates are on the order of 30 percent; for the branded pharmaceuticals, her best estimate is around 12 percent.

Similarly, mail order is an increasingly important distribution channel. It now accounts for about 18 percent of total pharmaceutical sales. The prescription is usually for a 3-month supply, whereas from a pharmacy it is a 1-month supply. So, again, if one simply counts the number of prescriptions to estimate the price per prescription, either the mail order users will be missed, or the wrong number will be calculated, because the content of every prescription is dramatically different, by a factor of three in this case.

Following up on Danzon's example, Cutler added that, for these cases in which different strengths in dosages and formulations exist and substitutions across them occur, the more aggregated index should perform better than the more detailed version, provided quality adjustment is done properly. The aggregated index will take the total growth in pharmaceutical spending at the condition level and then ask how much of that growth is the result of new drugs and what is their net value. In this respect, moving to the larger pricing buckets is actually the right way to go and immensely important. The question then becomes: Is there any way to link across those buckets?

Cutler added that, no matter which way the index is constructed, the quality adjustment is needed. It may be simpler for a more aggregated index, because each exact formulation does not have to be dealt with individually. Anything that is truly a new good will raise a different problem that will be missed either way; that is a big remaining issue.

3.5. OUTCOMES AND QUALITY CHANGE

At several points during the workshop, participants made the point that, to monitor quality change in medical care for purposes of price measurement, accurate data on outcomes for treatments—defined in parallel with the expenditure categories—would be needed. Mark McClellan, of the Brookings Institution and formerly of CMS and the Food and Drug Administration, spoke about measuring treatment outcome in this context. Among participants, there seemed to be complete agreement that quality adjustment of price indexes for the satellite health care accounts is extremely important and also that it is very hard to do.

McClellan began by noting that many of the measurement problems faced in the construction of price indexes for health accounts are increasingly important elsewhere for people—payers, purchasers, consumer groups, providers, and policy makers—concerned with where the health care system is going. The work that is going on in parallel creates an opportunity for collaboration on doing a better job of measuring outcomes and on putting more of a focus on value in the processes of health care decision making and policy.

During open discussion, Linda Bilheimer (National Center for Health Statistics) asked what policy makers are looking for in terms of measures of outcomes. McClellan responded that it depends on whom and when you ask. If it is a briefing before the Joint Economic Committee about where health care should be headed in the next 5 or 10 years, then topics related to value and to accounting for productivity changes arise. If it is a Congressional Budget Committee meeting on how to get Medicare physician payment problems addressed for next year, then the topics are about the price and the policy changes that could affect nominal budget spending.

As work moves forward in this area, one of the side benefits will be a better awareness among the general public and policy makers about the distinction in these kinds of questions. When people are asked today about what is wrong with health care, their response is increasing costs, and they equate that with prices going up—the premiums that they pay on their insurance plan and so forth. Even though they are individually quite satisfied with the care they are getting and perhaps reluctant to see major changes in health care policies that could directly affect their care, there is less recognition of the broader questions of how policies are affecting value in health care and what people are getting for what they spend.

Among the many challenges with measuring outcomes and accounting for them in indexes is sorting out the impact of health care on health, which can be hard to isolate, especially at the patient level; so many factors influence health. Also, there are few standard quality or outcome measures established for many aspects of health care. The trend has been to start with narrow pieces of the picture—such as a look at a specific disease—and try to expand that over time as data and technical expertise get better. McClellan reminded the audience that there is still a long way to go.

Next, McClellan reiterated the point made throughout the workshop about the measurement problems created by the existence of multiple chronic conditions. He noted that it is getting more and more difficult to isolate diseases that coexist in individuals. The vast majority of Medicare spending now is on people who have multiple chronic conditions. For these patients, health professionals are increasingly realizing that focusing on one particular disease and its treatment leads to real missed opportunities to improve the coordination and results of care. Accordingly, efforts are being made to cut across disease areas with prescriptions for behavioral changes and medications, compliance systems, and the like that are not easy to attribute in patients with multiple conditions or any particular disease.

Even when this kind of effort is made, however, McClellan agreed that it was hard to sort out how much of a given health status effect is due to the medical treatment and how much to other factors. Health is clearly improving over time, although at different rates for different kinds of disease treatments, and these trends reflect changes not only in medical technology, but also in biomedical knowledge that affects behavior, as well as nonmedical factors that are not measured as medical care in the economy. Wellness expenditures are a growing industry, and food improvement, education, socioeconomic status, and environmental exposures certainly affect health and are important determinants as well.

For all of these reasons, McClellan pointed out, there has not been a lot of practical application of outcome measures to ongoing health care policy aimed at improving the value of what society is getting for its spending. Instead, much of the focus has tended to be on process-of-care measures, for which it is easier to conclude with a reasonable level of confidence, from clinical studies and expert opinion, that using certain kinds of treatment for particular conditions or combinations of conditions leads to better results. As an aside, he noted that the results have not been particularly encouraging about whether the health care system is reliably delivering quality care. Nor have the these methods been very useful as surrogates for outcome measures; they have tended to focus on specific aspects of care and do not capture most of the things that consumers or even providers need to know in their decisions about health care.

On the positive side, McClellan reported that many efforts are under way to change the way information about the health care system is processed. This is where McClellan sees some parallels and some opportunities for collaboration between the kinds of people who attended the workshop and those who are working in such areas as quality improvement, payment reform, benefit redesign, and the like. Among the interested parties are provider groups that have been struggling with the traditional ways of paying for programs like Medicare based on volume and intensity, in which the final common pathway to address rising spending is to squeeze down prices; this, McClellan stated, is not working very well in terms of promoting quality and value or even long-term cost savings, whether it is health plans or the employers who use them, who want more accountability for what they are getting for their spending. There are also interested consumer groups, like Consumers Union, that are now engaged in initiatives to make health care much more like choosing appliances and cars; they want to see information about providers and health plans, just like those in *Consumer Reports* for these other areas.

CMS's Hospital Compare database is another example of how this work on quality measurement—and not only processes of care, but also outcomes and satisfaction measures—is progressing. The Hospital Compare site was implemented several years ago; since then it has expanded and now includes several outcome measures. There are now CMS reports on Hospital Compare for 30-day mortality from acute myocardial infarction and 30-day mortality from heart failure. McClellan expressed the hope that more will be coming soon in the area of surgical outcome measures and a range of other survival measures.

McClellan reported that a final area for which CMS is beginning to expand measures available on Hospital Compare—very much related to outcomes—involves standardized patient satisfaction measures. He said that surveyed patients generally respond favorably when asked whether they received satisfactory care. However, he continued, more detailed data providing a deeper understanding of relationships with providers and doctors and nurses, of how information about the condition was communicated, and of how patients felt about particular aspects of the treatments they received can all be very helpful. This turns out to be especially true for patients with chronic diseases, who often have a good idea of what is working to keep their conditions under control. Gail Wilensky commented that it is very helpful that people are being pushed to understand better, with tools like the Hospital Compare data or other data that are becoming available, that there are different ways to measure outcomes. She also noted that questions have to be framed very carefully. For example, when individuals are asked what is important to them about health care costs or health care prices, they typically think only in terms of what they are paying out-of-pocket—that is their working definition.

McClellan also mentioned that other parallel efforts are under way for physician care, nursing home care, and pharmacy care, but they are not as far along as the Hospital Compare data. They all follow a similar general model, starting with some process-of-care measures, then push toward looking at outcomes, patient satisfaction, and other aspects related to outcomes. With this increasing emphasis focusing on value—and not just volume and intensity and prices in health care reform—there have been a number of efforts bringing together provider groups, payers, purchasers, and consumer groups of health care. One purpose has been to get consensus behind methods to measure the quality of outcomes and costs, and to do it more at the episode or patient level, rather than just in a particular silo (such as hospitals, physicians' offices, etc.) of care. McClellan described the typical process as involving a number of organizations that become involved in developing the technical details of what a quality measure, whether it is process or outcome or satisfaction, might look like. Next, a process is coordinated by the National Quality Forum, a congressionally recognized nonprofit organization, to try to get a consensus endorsement behind particular measures.

These processes by themselves do not do anything to get the quality measures into use in practice. According to McClellan, there have been a number of collaborative activities developed over the last few years to do that. Most of them are in the form of quality alliances or hospital quality alliances; these are instrumental in creating a consensus behind the measures of care that are used in such systems as Hospital Compare. One of these is a group called AQA, formerly Ambulatory Care Quality Alliance, which is concerned mainly about physician and ambulatory care quality measurement and is behind some of the efforts by Medicare and private payers to put more emphasis on quality and payment reporting. On the ambulatory side, an organization called Pharm Quality Alliance is working toward some similar goals.

A group called the Quality Alliance Steering Committee has been charged with trying to help these groups work together, to collaborate in this effort to get more consistent and common metrics out of the nation's pluralistic health care system. The focus of the steering committee is to ensure that measures being developed in each of these silos, as McClellan described the various areas, are not only harmonized, but also on a track focusing on overall pictures of quality and cost at the patient level, or at least at the episode level, and on ways to get synergistic benefits from using data from multiple sources together. One of the challenges in developing and using these measures is that any health plan, even Medicare, does not get a complete picture of care quality at the level of providers and their treatment of diseases or other conditions, or even at the patient level. Much of the focus on these collaborative efforts has been on trying to harmonize the different measures that various health plans, Medicare, or employers are using that would facilitate an aggregated approach that provides a more complete picture of quality and cost of care.

3.6. DATA NEEDS FOR PRICE MEASUREMENT, TRACKING OUTCOMES, AND QUALITY ADJUSTMENT

Mark McClellan's presentation also touched on some overarching data issues for measuring quality change; many of these parallel points were made during the discussions on medical care expenditures. McClellan stated that the aggregated data—compiled at the level of the provider, the health plan, or the treatment of similar patients—are what matters, not the fact that a specific patient was treated by a specific doctor with certain results. If participation processes and data collection for the health care system were carried out with some consistency, it would be possible to perform complex analyses (e.g., multivariate regressions) and to produce relatively sophisticated measures in a distributed data system. Furthermore, he pointed out that if a truly electronic health care system were created, it would have much more analytic value—not necessarily in terms of data volume but through development of consistent rules and standards being applied that would enable researchers to use it much more effectively.

Danzon commented that there is a huge amount of data that are already collected by the pharmaceutical companies on comparative effectiveness of new technologies versus old technologies or new drugs versus old drugs that they have to collect in order to make their case for reimbursement in many foreign countries and increasingly with health plans in the United States. Many of the data are collected as part of clinical trials. With that comes limitations, but the data would provide some evidence about new technology versus old. In an efficient system, BLS and BEA would be able to take advantage of the millions of dollars spent collecting these outcomes data as part of this exercise.

Newhouse pointed out that McClellan actually did a paper 10 years ago or so on heart attacks in which clinical trial data were used to break down the components of improvements and outcomes and attributes them to changes in specific aspects of treatment. McClellan pointed out that data could continue to be used in that way; however, in terms of actual health care delivery, there is a big gap between how well the technologies could be used and how they are used in practice. In most cases, even when there is a big medical breakthrough, it does not appear in the data from one year to the next.

He cited the example of beta blockers used in heart attack patients—a treatment that, beginning in the late 1970s, was used to substantially improve survival of patients after acute myocardial infarction. In the 1980s it started to be tracked in a few limited settings, and in the 1990s it became part of a routine quality measure that was part of Hospital Compare for a while. But it took three decades to get from the time when the studies were done to when there was complete use and practice of the procedure. So these kinds of data, in conjunction with trend information on the use of different technologies, can certainly be useful. But they are not going to provide a complete picture. The only other caution he added is that the quality adjustments could be significant; the measures on outcomes are going to get better over time, but they are almost certainly going to be different from year to year—it is very hard to maintain consistency over longer time periods.

Triplett pointed out that there have still been only a small number of disease-based studies in the economics literature—the heart attack study (Cutler et al., 1998), the cataract study (Shapiro et al., 2001), the mental health study (Berndt et al., 2001), and a couple of others. However, outcome measures were not used in any of these. The heart attack study used mortality, but this was not a full outcome measure—it is a lower bound, because it does not pick up morbidity effects. For the cataract and mental health studies, there were no direct outcome measures either (for cataracts, the researchers contended that they had estimated a lower bound). Triplett concluded that there is not that much low-hanging fruit from the literature to pick up. Despite agreement in the past couple of decades that outcome measures are needed to conduct cost-effectiveness studies—which everyone agreed are important, as they are being used for health care planning as well as in research—the medical literature on the topic is still not extensive, and that limits what economists can do.

McClellan responded that the technical ability to do something about this, to come up with some more reasonable and more complete outcome measures, though still very incomplete in terms of everything people might care about, has gotten a lot better. There is currently clearer policy agreement that just focusing on volume and price restrictions is not going to be enough. That said, McClellan agreed that there is still a long way to go. But, as more work gets done in this area, one of the side benefits is going to be a better awareness among the general public and policy makers about the distinction between cost and price and productivity questions.

Some of the movement in the direction of quality measurement is also being driven by policy and legal pressures. McClellan cited a recent settlement in which the New York attorney general required transparency and the use of nationally recognized quality standards from major health insurers who have been trying to use measures of quality and costs of care as either conditions in their contracting with providers or as factors that influence the structure of their benefits (e.g., setting lower copays and perhaps paying the physicians and hospitals according to performance based on whatever measure each health plan or each employer came up with). The point here is that these metrics need to be based on both quality and cost, and they need to have a more comprehensive and consistent picture than a health plan is likely to be able to get on its own with the limited number of patients that it covers. There has been an added push for these efforts to do network or aggregated approaches to quality and cost measurement.

The practical relevance of this to the work that is being done in health accounts and price indexes is still a way off. However, McClellan pointed out that some of these broader measures are in the early stages of being constructed and made available. There is a broad national public-private roadmap planning effort to move from data that are based just on claims to data that include what might be called clinically enriched electronic information, like lab results and increasingly sophisticated information from electronic records or personal health records. There is a parallel between the kind of work that is going on here and the kind of work that is going on in the health accounts area.

McClellan expressed his concern that the initiatives he is involved with, as well as the overlapping health accounting programs, move forward as effectively as possible from a policy reform standpoint; this will require efficient use of data. Ideally, a virtuous cycle could be created in which, with more and better information available on outcomes and costs of care and therefore on the value of care, there will be a movement toward payment and benefit system designs that reward and support better value and clearer evidence about what actually works.

McClellan concluded his remarks by suggesting that some ongoing involvement of the BEA and the National Academies on work that is happening in these quality measurement, value measurement, and quality improvement efforts would make a lot of sense. The measurement goals across the various interests are similar; the only difference is that BEA has to focus on the national accounts as opposed to the actual impact on delivery of care.

4

Summary, Perspective, and Prospects for Moving Forward

At the end of the day, members of the steering committee (and others) noted several areas of topical guidance for which much agreement among workshop participants was demonstrated; several of these areas include actions for the Bureau of Economic Analysis (BEA) and other agencies to consider. Participants supported the idea that BEA's satellite account, and any subsequent revision to the medical care component of the national income and product accounts (NIPAs), should not involve a revision of scope. Rather, the new framework should stay within the market-oriented boundaries of the national accounts and explicitly extend to a nonmarket component. BEA's new work is primarily intended to provide an alternative way of measuring medical care prices and quantities, and the agency's motivation is to improve measurement within the existing NIPAs. BEA presenters acknowledged that the work could eventually affect overall measured rates of inflation and productivity and, in turn, real gross domestic product growth rates.

Barbara Fraumeni, Jack Triplett, and others were pleased that BEA is pushing forward with work on the full structure of the accounts—both the expenditures and industry sides—and were reassured to learn that the agency had a strategy for updating the industry accounts. Because the medical care sector is complicated, it is important that work progresses in parallel with expenditure-side revisions from the start.

Workshop participants also agreed that the methodological focus on episodes of treatment as the unit of measurement was the right one. They supported the notion that treatments could be meaningfully priced, in the sense that treatment is the service concept that consumers (patients) demand, and can be coordinated with current national income and product accounting processes.

Participants demonstrated different preferences regarding specifically how to track expenditures and output (and eventually outcomes) associated with medical treatments. Whether by encounter, by episode, or by person, all agreed that the appropriate concept depends a great deal on the specific application. For example, for price index work, a person-based approach may not be as appropriate as an episode-based approach. If the goal is to broadly compare costs and health improvements within a given disease on a micro level, as is done in cost-effectiveness studies and decision analysis, a person-based regression approach might be right. BEA participants noted that more research is needed on the virtues and limitations of various methods.

Participants embraced BEA's two-stage strategy for implementing the satellite account that involves first getting expenditures classified, then moving on to quality change issues. There was general acknowledgment that, because such information serves as a building block for many kinds of health data systems, creating new ways of organizing and tracking health care expenditures is an immediate priority. This work should prove useful for both the experimental health accounting and national income and product accounting purposes, as well as for price and productivity measurement. Once the nominal flows for the sector have been figured out correctly, BEA can then move on to estimating improved disease-based price indexes.

Although the plan is to defer integration of some aspects of price work until a greater consensus about methods emerges, participants emphasized the importance of thinking about quality change from the beginning. Ideally, the satellite account should move forward in coordination with work proceeding elsewhere in the health economics literature, which implies focusing on outcome trends. Ultimately, moving beyond measurement only of medical care inputs will allow researchers to gain greater insights into traditional economic issues, such as productivity growth and quality change in medical care. This underscores the importance of drawing on expertise from medical researchers to help evaluate trends in treatment outcomes.

Participants also agreed that medical care deflators or indexes should reflect price changes associated with changes in inputs—for example, moving from surgical to drug treatment or inpatient to outpatient settings; ideally, they would also be able to assess whether quality has improved, deteriorated, or remained constant. If quality changes, for better or worse, that should be reflected in the calculation of real output. Both BEA and the Bureau of Labor Statistics (BLS) presented research showing progress in this area but noted that the capability to adjust price indexes to reflect trends in outcomes is still a long way off, and the agencies would be relying heavily on the academic community to point the way.

Several participants suggested that BEA (and BLS) should do more to document their plans and progress. Requests were made for two papers from BEA: one that discusses the accounting system and proposed changes to it; and another that describes what can and cannot be done now and what the agency would like to tackle in the future—namely, the quality change issue. One of the most difficult issues that arises in health accounting and price indexing work is comorbidity; when patients require medical care for multiple conditions, the task of assigning expenditures to predefined categories accurately becomes much more complicated. This will be a key topic in these papers.

Data needs for advancing health accounting were identified at several points throughout the day. Dale Jorgenson emphasized the need to consider a wide range of possible data sources to underpin the satellite health care account. His suggestion to BEA and BLS was that they work out a way of coordinating their data infrastructures as efficiently as possible. He made the point that a solution will need to be found for combining data on providers' prices with the information collected from claims and suggested that kind of work be put on the table for BEA. Ana Aizcorbe agreed that exploiting multiple data sources was a key task, but she noted a number of difficulties, among them the barriers to linking provider data on expenditures with data on patients. A lot could be done with claims data for the insured population if, for no other reason, because of their enormous size and coverage. BEA staff agreed that using the Medical Expenditure Panel Survey (MEPS) as the backbone of the data infrastructure, and then claims information in a supplemental role wherever gaps appear, was a reasonable strategy. Many participants noted the importance of anticipating how alternative strategies will play out once the research program is in full swing.

To monitor quality change in medical care for purposes of price measurement, accurate data on outcomes for treatments—defined in parallel with the expenditure categories—would be needed. Mark McClellan spoke about measuring treatment outcomes in this context. Among participants, there seemed to be complete agreement that quality adjustment of price indexes for the satellite health care accounts is extremely important, and also that it is very hard to do.

Gail Wilensky noted two specific types of data needs that will need to be combined: (1) individual observations in a national probability sample (e.g., MEPS), which provides good measures for conditions that occur frequently enough in a relatively small sample and for which a good reliable statistical basis can be formed, and (2) registries or other specialized data for those rarer conditions that would otherwise require very large data sets to provide that information.

In closing comments, Wilensky summarized the role of BEA's program, noting that many of the key tasks will involve long-term research commitments. Because of this, using a mechanism such as a satellite account—in which experimental strategies can be explored while not interfering with the workings of the regular national accounts—offers a good strategy for balancing the need for additional information with the need to maintain methodologies that have a proven historical record.

References

Berndt, Ernst R., Susan H. Busch, and Richard G. Frank. (1998). *Price Indexes for Acute Phase Treatment of Depression*. (NBER Working Paper No. W6799.) Cambridge, MA: National Bureau of Economic Research.

Cutler, David M., Mark McClellan, Joseph P. Newhouse, and Dahlia Remler. (1998). Are medical prices declining? Evidence for heart attack treatments. *Quarterly Journal of Economics, 113*(4), 991-1024.

National Research Council. (2005). *Beyond the Market: Designing Nonmarket Accounts for the United States*. Panel to Study the Design of Nonmarket Accounts, Katharine G. Abraham and Christopher Mackie, Eds. Washington, DC: The National Academies Press.

Rice, Dorothy P. (1966). *Estimating the Cost of Illness*. (Health Economics Series, No. 6, DHEW Publication No. (PHS)947-6.) Rockville, MD: U.S. Department of Health, Education, and Welfare.

Rice, Dorothy P., Thomas A. Hodgson, and Andrea N. Kopstein. (1985). The economic cost of illness: A replication and update. *Health Care Financing Review, 6*(1), 61-80.

Scitovsky, Anne A. (1967). Changes in the costs of treatment of selected illnesses, 1951-1995. *American Economic Review, 57*(5), 1182-1195.

Shapiro, Irving, Matthew D. Shapiro, and David W. Wilcox. (2001). Measuring the value of cataract surgery. In *Medical Care Output and Productivity*, David M. Cutler and Ernst R. Berndt, Eds. Chicago: University of Chicago Press.

Sing, Merrile, Jessica Banthin, and Thomas M. Selden. (2006). Reconciling medical expenditure estimates from the Medical Expenditure Panel Survey and the national health accounts, 2002. *Health Care Financing Review, 28*(1), 25-40.

Appendix A

Summary Statistics from the Medical CPI and U.S. Medical Expenditures Panel Survey

The tables in this appendix were distributed in a handout at the workshop by participants from the Bureau of Labor Statistics. Table A.1 lists the number of people in each conditions field from the associated Medical Expenditure Panel Survey (MEPS) file; these are national estimates, so person weights are being used for cases in which at least one diagnosis is present for these diseases. For example, in 1998, 25 million people had at least one diagnosis of infectious diseases. Table A.2 gives the nominal expenditures that are in the MEPS event files by disease, by year. For example, $1.61 billion was spent on physicians for infectious and parasitic diseases in 1998. Table A.3 gives utilization or quantities used to adjust the experimental disease index. For example, in 1998, for infectious and parasitic diseases, there was an average of 1.2 physician visits for that year. Table A.4 details, by source, expenditures based on the consumer price index strata for various services. Table A.5 gives the cumulative growth rate from January 1999 to December 2004 by disease for various indexes.

TABLE A.1 Number of Diagnoses by Major Disease and Year, 1998-2004 (in millions)

Disease	1998	1999	2000	2001	2002	2003	2004
Infectious diseases	25.1	23.8	24.5	26.2	26.1	26.0	23.9
Neoplasms	17.2	16.9	17.2	18.9	20.7	20.6	20.1
Endocrine, nutritional, and related diseases	47.1	50.2	55.0	60.8	64.7	67.7	75.6
Diseases of the blood	3.1	3.3	3.9	4.2	4.2	4.1	4.2
Mental disorders	40.7	38.2	39.8	45.7	54.5	56.0	59.7
Diseases of the nervous system	85.5	79.1	76.9	81.7	82.6	86.6	88.2
Diseases of the circulatory system	65.7	65.1	68.8	72.4	80.0	83.6	87.5
Diseases of the respiratory system	175.6	172.7	168.9	183.2	179.1	184.4	177.4
Diseases of the digestive system	79.1	82.1	82.7	83.4	90.4	93.8	92.2
Diseases of the genitourinary system	34.7	35.3	38.0	40.8	41.3	41.8	41.3
Complications of pregnancy	13.7	14.6	16.9	18.4	18.0	19.0	18.8
Diseases of the skin	27.4	25.8	28.2	31.4	31.6	30.9	29.2
Diseases of the musculoskeletal system	75.9	75.8	76.4	86.3	96.6	99.6	102.6
Congenital anomalies	2.3	1.6	1.6	1.7	1.7	1.8	1.9
Certain conditions in the perinatal period	0.4	0.5	0.8	0.8	0.9	1.1	0.9
Injury and poisoning	64.3	60.1	60.8	64.7	66.1	68.0	68.5
Other conditions	64.2	66.6	71.3	79.2	81.7	83.4	83.7

SOURCE: Workshop presentation by Ralph Bradley.

TABLE A.2 National Expenditures for Various Medical Services by Major Disease and Year, 1998-2004 (in billions of dollars)

Year	Disease	Physicians	Outpatient Services	Emergency Room	Pharmaceutical	Inpatient Hospital	Total	Cumulative Growth in Totals
1998	Infectious and parasitic diseases	1.61	0.56	0.44	1.92	2.81	7.33	1.00
1999	Infectious and parasitic diseases	2.32	0.52	0.38	2.21	5.72	11.16	1.52
2000	Infectious and parasitic diseases	2.07	0.54	0.25	1.74	4.56	9.16	1.25
2001	Infectious and parasitic diseases	2.46	0.50	0.55	2.58	2.19	8.27	1.13
2002	Infectious and parasitic diseases	2.62	0.81	0.54	4.21	8.21	16.39	2.23
2003	Infectious and parasitic diseases	3.15	0.88	0.47	4.69	2.55	11.74	1.60
2004	Infectious and parasitic diseases	3.19	0.85	0.65	5.41	4.40	14.50	1.98
1998	Neoplasms	9.08	8.01	0.18	1.51	22.54	41.32	1.00
1999	Neoplasms	8.59	5.70	0.24	1.36	17.86	33.75	0.82
2000	Neoplasms	9.95	7.22	0.22	1.16	23.78	42.33	1.02
2001	Neoplasms	14.61	8.76	0.31	1.82	23.81	49.31	1.19
2002	Neoplasms	15.03	9.78	0.41	1.68	25.79	52.69	1.28
2003	Neoplasms	13.32	15.41	0.36	1.77	23.07	53.93	1.31
2004	Neoplasms	22.85	11.62	0.49	2.09	30.19	67.25	1.63
1998	Endocrine, nutritional, and metabolic diseases and immunity disorders	7.63	1.49	0.39	10.85	9.39	29.75	1.00
1999	Endocrine, nutritional, and metabolic diseases and immunity disorders	7.25	1.04	0.35	13.89	7.86	30.38	1.02
2000	Endocrine, nutritional, and metabolic diseases and immunity disorders	9.04	1.97	0.65	16.42	6.77	34.85	1.17
2001	Endocrine, nutritional, and metabolic diseases and immunity disorders	9.42	4.19	0.93	23.18	9.55	47.27	1.59

continued

TABLE A.2 Continued

Year	Disease	Physicians	Outpatient Services	Emergency Room	Pharma-ceutical	Inpatient Hospital	Total	Cumulative Growth in Totals
2002	Endocrine, nutritional, and metabolic diseases and immunity disorders	12.45	3.37	0.65	24.19	9.89	50.55	1.70
2003	Endocrine, nutritional, and metabolic diseases and immunity disorders	13.13	2.52	0.95	29.72	11.10	57.43	1.93
2004	Endocrine, nutritional, and metabolic diseases and immunity disorders	16.24	2.86	0.87	33.72	14.45	68.14	2.29
1998	Diseases of the blood and blood-forming organs	0.61	0.24	0.05	0.21	1.16	2.27	1.00
1999	Diseases of the blood and blood-forming organs	0.47	0.21	0.03	0.31	2.27	3.29	1.45
2000	Diseases of the blood and blood-forming organs	0.81	0.32	0.06	0.31	1.09	2.60	1.15
2001	Diseases of the blood and blood-forming organs	1.26	0.76	0.05	0.55	2.38	5.00	2.21
2002	Diseases of the blood and blood-forming organs	1.50	1.03	0.15	0.45	1.96	5.09	2.24
2003	Diseases of the blood and blood-forming organs	1.40	0.26	0.08	0.64	1.80	4.18	1.85
2004	Diseases of the blood and blood-forming organs	1.38	0.84	0.20	0.54	2.32	5.27	2.33
1998	Mental disorders	9.26	3.51	0.25	9.24	12.08	34.35	1.00
1999	Mental disorders	8.01	2.27	0.22	8.78	17.18	36.45	1.06
2000	Mental disorders	9.48	1.94	0.27	13.77	7.68	33.14	0.96
2001	Mental disorders	11.23	1.19	0.57	15.50	10.66	39.13	1.14
2002	Mental disorders	14.89	1.73	0.96	17.31	12.58	47.47	1.38
2003	Mental disorders	13.60	2.05	0.82	20.19	9.95	46.61	1.36
2004	Mental disorders	15.63	1.61	0.73	22.94	10.64	51.55	1.50
1998	Diseases of the nervous system and sense organs	13.04	7.33	1.33	5.20	9.09	35.99	1.00
1999	Diseases of the nervous system and sense organs	12.05	5.59	1.11	5.83	6.19	30.76	0.85
2000	Diseases of the nervous system and sense organs	12.65	5.38	1.16	6.35	9.94	35.49	0.99

2001	Diseases of the nervous system and sense organs	15.32	6.95	1.95	7.42	9.60	41.24	1.15
2002	Diseases of the nervous system and sense organs	20.22	8.43	1.84	9.07	10.72	50.29	1.40
2003	Diseases of the nervous system and sense organs	20.67	8.55	1.90	10.20	51.99	93.31	2.59
2004	Diseases of the nervous system and sense organs	22.86	7.63	2.37	10.72	11.31	54.90	1.53
1998	Diseases of the circulatory system	11.42	5.69	2.62	16.02	63.23	98.98	1.00
1999	Diseases of the circulatory system	12.34	4.83	1.93	19.24	66.20	104.55	1.06
2000	Diseases of the circulatory system	13.85	6.50	2.96	19.76	71.51	114.57	1.16
2001	Diseases of the circulatory system	15.87	7.80	3.73	22.97	64.19	114.57	1.16
2002	Diseases of the circulatory system	20.34	7.85	4.29	25.48	67.99	125.95	1.27
2003	Diseases of the circulatory system	18.72	10.61	4.71	29.85	79.54	143.43	1.45
2004	Diseases of the circulatory system	22.50	10.71	6.38	32.57	91.06	163.21	1.65
1998	Diseases of the respiratory system	10.75	2.35	1.74	9.21	27.63	51.67	1.00
1999	Diseases of the respiratory system	10.07	2.52	2.16	11.33	22.72	48.80	0.94
2000	Diseases of the respiratory system	10.31	2.26	2.22	11.28	26.35	52.43	1.01
2001	Diseases of the respiratory system	13.37	3.10	2.78	16.76	29.37	65.37	1.27
2002	Diseases of the respiratory system	15.52	4.71	3.13	19.40	24.79	67.54	1.31
2003	Diseases of the respiratory system	14.57	3.43	3.40	20.25	26.97	68.62	1.33
2004	Diseases of the respiratory system	13.93	3.31	3.64	19.88	38.33	79.09	1.53
1998	Diseases of the digestive system	4.42	4.70	1.39	5.10	20.48	36.09	1.00
1999	Diseases of the digestive system	4.84	4.59	2.05	5.82	18.24	35.54	0.98
2000	Diseases of the digestive system	5.33	4.68	1.99	6.56	21.07	39.62	1.10
2001	Diseases of the digestive system	5.91	6.37	2.31	9.79	23.18	47.56	1.32
2002	Diseases of the digestive system	7.53	6.79	2.53	11.67	29.79	58.32	1.62
2003	Diseases of the digestive system	8.19	8.17	3.46	15.51	24.52	59.85	1.66
2004	Diseases of the digestive system	10.90	9.46	3.75	16.26	43.23	83.59	2.32

continued

TABLE A.2 Continued

Year	Disease	Physicians	Outpatient Services	Emergency Room	Pharma-ceutical	Inpatient Hospital	Total	Cumulative Growth in Totals
1998	Diseases of the genitourinary system	5.94	4.66	1.05	3.39	9.17	24.21	1.00
1999	Diseases of the genitourinary system	5.85	3.56	0.63	3.74	9.89	23.68	0.98
2000	Diseases of the genitourinary system	6.26	5.18	1.41	3.76	11.60	28.21	1.17
2001	Diseases of the genitourinary system	9.14	7.69	1.61	4.44	10.25	33.13	1.37
2002	Diseases of the genitourinary system	12.02	9.28	2.52	5.03	10.96	39.82	1.64
2003	Diseases of the genitourinary system	11.71	8.88	2.27	6.54	17.85	47.26	1.95
2004	Diseases of the genitourinary system	14.85	10.24	2.68	5.80	16.73	50.30	2.08
1998	Complications of pregnancy, childbirth, and the puerperium	3.81	1.34	0.36	0.78	15.09	21.39	1.00
1999	Complications of pregnancy, childbirth, and the puerperium	4.60	0.93	0.19	1.15	15.81	22.69	1.06
2000	Complications of pregnancy, childbirth, and the puerperium	5.35	0.84	0.76	1.20	16.79	24.95	1.17
2001	Complications of pregnancy, childbirth, and the puerperium	5.59	1.72	0.53	1.46	18.28	27.58	1.29
2002	Complications of pregnancy, childbirth, and the puerperium	5.94	0.97	0.63	1.53	24.09	33.16	1.55
2003	Complications of pregnancy, childbirth, and the puerperium	6.14	1.39	0.57	2.19	24.38	34.67	1.62
2004	Complications of pregnancy, childbirth, and the puerperium	6.45	2.04	0.61	2.16	23.93	35.20	1.65
1998	Diseases of the skin and subcutaneous tissue	3.41	1.13	0.23	1.82	1.96	8.56	1.00
1999	Diseases of the skin and subcutaneous tissue	3.27	0.96	0.18	2.13	2.15	8.69	1.02
2000	Diseases of the skin and subcutaneous tissue	3.65	1.10	0.21	2.16	4.71	11.83	1.38

2001	Diseases of the skin and subcutaneous tissue	4.55	1.47	0.62	2.83	2.66	12.12	1.42
2002	Diseases of the skin and subcutaneous tissue	5.37	1.48	0.40	2.85	4.50	14.59	1.71
2003	Diseases of the skin and subcutaneous tissue	5.21	1.24	0.57	3.23	4.78	15.02	1.76
2004	Diseases of the skin and subcutaneous tissue	5.56	1.74	0.50	3.15	6.94	17.89	2.09
1998	Diseases of the musculoskeletal system and connective tissue	16.78	6.55	1.04	5.45	15.10	44.93	1.00
1999	Diseases of the musculoskeletal system and connective tissue	15.30	5.75	0.64	7.40	16.92	46.02	1.02
2000	Diseases of the musculoskeletal system and connective tissue	20.15	5.41	0.84	8.55	15.50	50.45	1.12
2001	Diseases of the musculoskeletal system and connective tissue	24.60	7.23	1.33	12.21	20.53	65.89	1.47
2002	Diseases of the musculoskeletal system and connective tissue	26.84	9.33	1.87	13.63	19.59	71.26	1.59
2003	Diseases of the musculoskeletal system and connective tissue	30.61	11.07	1.23	17.22	21.90	82.03	1.83
2004	Diseases of the musculoskeletal system and connective tissue	34.10	12.51	2.60	17.82	23.54	90.57	2.02
1998	Congenital anomalies	0.41	0.41	0.04	0.09	2.32	3.26	1.00
1999	Congenital anomalies	0.61	0.27	0.02	0.07	3.19	4.17	1.28
2000	Congenital anomalies	0.54	0.51	0.01	0.07	1.00	2.13	0.65
2001	Congenital anomalies	0.68	0.51	0.15	0.15	1.53	3.02	0.92
2002	Congenital anomalies	0.78	0.92	0.02	0.19	2.08	3.98	1.22
2003	Congenital anomalies	0.85	0.49	0.04	0.15	4.51	6.03	1.85
2004	Congenital anomalies	1.20	0.88	0.02	0.22	1.49	3.81	1.17

continued

TABLE A.2 Continued

Year	Disease	Physicians	Outpatient Services	Emergency Room	Pharmaceutical	Inpatient Hospital	Total	Cumulative Growth in Totals
1998	Injury and poisoning	11.50	6.47	6.42	1.08	19.56	45.03	1.00
1999	Injury and poisoning	12.89	6.53	6.95	1.56	23.79	51.73	1.15
2000	Injury and poisoning	13.59	5.98	6.14	1.74	19.60	47.05	1.04
2001	Injury and poisoning	14.93	5.99	8.09	1.85	24.21	55.06	1.22
2002	Injury and poisoning	16.27	5.31	9.04	2.26	27.81	60.69	1.35
2003	Injury and poisoning	16.92	7.91	9.15	2.38	59.44	95.80	2.13
2004	Injury and poisoning	19.94	8.85	11.37	2.20	25.94	68.29	1.52
1998	Other conditions	5.04	2.35	0.80	5.87	7.39	21.44	1.00
1999	Other conditions	5.91	1.66	0.87	7.47	4.22	20.13	0.94
2000	Other conditions	6.27	2.14	0.76	8.74	5.81	23.73	1.11
2001	Other conditions	8.15	2.06	1.05	11.45	5.26	27.97	1.30
2002	Other conditions	9.85	3.01	1.40	13.15	8.72	36.12	1.68
2003	Other conditions	9.69	3.45	1.48	14.48	19.56	48.66	2.27
2004	Other conditions	11.48	3.36	2.03	16.64	10.65	44.16	2.06
1998	No diagnosis	1.53	0.15	0.02	0.12	0.01	1.82	1.00
1999	No diagnosis	1.30	0.17	0.01	0.12	0.02	1.62	0.89
2000	No diagnosis	1.28	0.10	0.01	0.14	0.09	1.62	0.89
2001	No diagnosis	1.56	0.23	0.05	0.10	0.01	1.94	1.06
2002	No diagnosis	1.89	0.24	0.08	0.18	0.02	2.41	1.32
2003	No diagnosis	1.62	0.29	0.02	0.14	0.00	2.07	1.13
2004	No diagnosis	2.12	0.21	0.08	0.33	0.01	2.75	1.51

SOURCE: Bureau of Labor Statistics handout at the workshop; see text.

TABLE A.3 Mean Utilization of Various Medical Services by Major Disease and Year, 1998-2004

Year	Disease	Mean Office Visits	Mean Hospital Admissions	Mean Hospital Number of Nights	Mean Emergency Room Visits	Mean Outpatient Visits	Mean Prescriptions
1998	Infectious and parasitic diseases	1.2236	0.0209	0.1202	0.0530	0.0945	0.9731
1999	Infectious and parasitic diseases	1.3810	0.0359	0.2287	0.0651	0.0811	1.0283
2000	Infectious and parasitic diseases	1.3001	0.0233	0.2649	0.0463	0.0768	0.9595
2001	Infectious and parasitic diseases	1.2851	0.0237	0.1137	0.0668	0.0776	0.9496
2002	Infectious and parasitic diseases	1.3524	0.0398	0.4376	0.0696	0.1035	0.9537
2003	Infectious and parasitic diseases	1.3855	0.0273	0.1645	0.0626	0.1048	1.0476
2004	Infectious and parasitic diseases	1.3689	0.0321	0.1709	0.0740	0.0963	1.0495
1998	Neoplasms	3.1064	0.1457	0.8734	0.0231	0.9312	0.5913
1999	Neoplasms	3.6210	0.1275	0.8367	0.0260	0.9735	0.6670
2000	Neoplasms	3.4116	0.1306	0.9531	0.0453	0.8232	0.5834
2001	Neoplasms	3.7468	0.1250	1.0338	0.0343	1.0669	0.7089
2002	Neoplasms	3.5347	0.1180	0.9139	0.0417	0.9083	0.5469
2003	Neoplasms	3.2690	0.1092	0.6854	0.0367	1.1519	0.5091
2004	Neoplasms	4.0979	0.1048	0.6971	0.0393	0.9771	0.5722
1998	Endocrine, nutritional, and metabolic diseases and immunity disorders	1.7589	0.0242	0.1709	0.0249	0.1176	2.2373
1999	Endocrine, nutritional, and metabolic diseases and immunity disorders	1.8338	0.0280	0.1845	0.0219	0.1122	2.5034
2000	Endocrine, nutritional, and metabolic diseases and immunity disorders	1.8667	0.0262	0.1399	0.0231	0.1494	2.4908
2001	Endocrine, nutritional, and metabolic diseases and immunity disorders	1.6095	0.0244	0.1766	0.0245	0.1963	2.4307

continued

TABLE A.3 Continued

Year	Disease	Mean Office Visits	Mean Hospital Admissions	Mean Hospital Number of Nights	Mean Emergency Room Visits	Mean Outpatient Visits	Mean Prescriptions
2002	Endocrine, nutritional, and metabolic diseases and immunity disorders	1.7970	0.0250	0.1643	0.0262	0.1715	2.4305
2003	Endocrine, nutritional, and metabolic diseases and immunity disorders	1.8119	0.0255	0.1421	0.0259	0.1400	2.5238
2004	Endocrine, nutritional, and metabolic diseases and immunity disorders	1.8725	0.0225	0.2115	0.0227	0.1296	2.5136
1998	Diseases of the blood and blood-forming organs	3.4285	0.0643	0.4220	0.0437	0.3589	0.9949
1999	Diseases of the blood and blood-forming organs	2.1867	0.0666	0.5243	0.0278	0.1970	1.1658
2000	Diseases of the blood and blood-forming organs	2.5232	0.0538	0.2253	0.0173	0.3310	1.3447
2001	Diseases of the blood and blood-forming organs	2.3217	0.1087	0.4886	0.0578	0.4512	1.1011
2002	Diseases of the blood and blood-forming organs	2.9677	0.1141	0.5096	0.0667	0.6263	0.9778
2003	Diseases of the blood and blood-forming organs	3.0383	0.0573	0.3197	0.0664	0.2180	0.9380
2004	Diseases of the blood and blood-forming organs	2.4521	0.1043	0.6139	0.0726	0.4862	1.0349
1998	Mental disorders	4.7332	0.0539	0.9543	0.0301	0.4362	1.8566
1999	Mental disorders	4.3933	0.0464	0.5197	0.0258	0.4570	1.9756
2000	Mental disorders	4.4950	0.0338	0.2949	0.0261	0.3297	2.0237
2001	Mental disorders	3.7914	0.0527	0.4110	0.0488	0.1782	2.0307

2002	Mental disorders	3.8819	0.0437	0.3906	0.0439	0.1755	2.0127
2003	Mental disorders	3.8347	0.0381	0.4189	0.0407	0.2123	2.0235
2004	Mental disorders	3.9862	0.0348	0.3994	0.0359	0.2175	2.0572
1998	Diseases of the nervous system and sense organs	1.9514	0.0163	0.1275	0.0643	0.1440	1.1255
1999	Diseases of the nervous system and sense organs	1.8573	0.0156	0.1018	0.0547	0.1055	1.1270
2000	Diseases of the nervous system and sense organs	1.8705	0.0189	0.1008	0.0627	0.1208	1.1418
2001	Diseases of the nervous system and sense organs	1.9397	0.0185	0.1169	0.0697	0.1460	1.1565
2002	Diseases of the nervous system and sense organs	2.1008	0.0199	0.1979	0.0718	0.1786	1.1258
2003	Diseases of the nervous system and sense organs	2.1749	0.0188	0.1941	0.0643	0.1582	1.1096
2004	Diseases of the nervous system and sense organs	2.1025	0.0167	0.1408	0.0631	0.1327	1.0828
1998	Diseases of the circulatory system	2.0754	0.0946	0.6922	0.0703	0.2546	2.5144
1999	Diseases of the circulatory system	2.0493	0.1079	0.6768	0.0662	0.2167	2.7369
2000	Diseases of the circulatory system	2.0463	0.1095	0.7434	0.0845	0.3516	2.6195
2001	Diseases of the circulatory system	2.0276	0.0985	0.6220	0.0859	0.2885	2.7506
2002	Diseases of the circulatory system	2.0976	0.0890	0.5522	0.0866	0.2412	2.6152
2003	Diseases of the circulatory system	2.0843	0.0886	0.5877	0.0811	0.2807	2.6643
2004	Diseases of the circulatory system	2.0546	0.0969	0.7219	0.0821	0.2410	2.7529
1998	Diseases of the respiratory system	1.3310	0.0302	0.2023	0.0613	0.0461	1.5380
1999	Diseases of the respiratory system	1.3178	0.0290	0.1739	0.0657	0.0467	1.5630
2000	Diseases of the respiratory system	1.2969	0.0304	0.2254	0.0581	0.0536	1.5892

continued

TABLE A.3 Continued

Year	Disease	Mean Office Visits	Mean Hospital Admissions	Mean Hospital Number of Nights	Mean Emergency Room Visits	Mean Outpatient Visits	Mean Prescriptions
2001	Diseases of the respiratory system	1.3657	0.0290	0.1992	0.0608	0.0549	1.6520
2002	Diseases of the respiratory system	1.4015	0.0279	0.1673	0.0690	0.0788	1.6418
2003	Diseases of the respiratory system	1.3170	0.0280	0.1669	0.0686	0.0565	1.6078
2004	Diseases of the respiratory system	1.3276	0.0327	0.2292	0.0734	0.0638	1.6519
1998	Diseases of the digestive system	1.0991	0.0749	0.3365	0.0820	0.1474	1.1945
1999	Diseases of the digestive system	0.9684	0.0632	0.3453	0.0785	0.1198	1.1997
2000	Diseases of the digestive system	0.9723	0.0618	0.3673	0.0854	0.1255	1.2256
2001	Diseases of the digestive system	0.9556	0.0629	0.3021	0.0890	0.1379	1.3280
2002	Diseases of the digestive system	0.9907	0.0629	0.3196	0.0897	0.1422	1.2872
2003	Diseases of the digestive system	1.0613	0.0586	0.2884	0.0940	0.1365	1.3331
2004	Diseases of the digestive system	0.9877	0.0642	0.3574	0.0912	0.1506	1.3425
1998	Diseases of the genitourinary system	1.9154	0.0626	0.2641	0.0759	0.3805	1.2889
1999	Diseases of the genitourinary system	1.5096	0.0459	0.2114	0.0664	0.1976	1.3634
2000	Diseases of the genitourinary system	1.5128	0.0476	0.2013	0.0687	0.4445	1.3171
2001	Diseases of the genitourinary system	1.8411	0.0479	0.1676	0.0770	0.5529	1.2981
2002	Diseases of the genitourinary system	2.1907	0.0504	0.2230	0.0912	0.5597	1.2770
2003	Diseases of the genitourinary system	2.1940	0.0460	0.2520	0.0925	0.4301	1.2796
2004	Diseases of the genitourinary system	2.2455	0.0531	0.2784	0.0965	0.4568	1.2822
1998	Complications of pregnancy, childbirth, and the puerperium	3.3424	0.3046	0.8987	0.0834	0.3021	0.9071
1999	Complications of pregnancy, childbirth, and the puerperium	3.2021	0.2663	0.7609	0.0488	0.1813	1.0745

2000	Complications of pregnancy, childbirth, and the puerperium	3.0388	0.2751	0.8858	0.0887	0.1562	1.0039
2001	Complications of pregnancy, childbirth, and the puerperium	2.7624	0.2422	0.8438	0.0724	0.2236	1.0668
2002	Complications of pregnancy, childbirth, and the puerperium	2.9531	0.2382	1.0157	0.0818	0.1556	1.0527
2003	Complications of pregnancy, childbirth, and the puerperium	2.9300	0.2447	0.9399	0.0853	0.1904	1.0800
2004	Complications of pregnancy, childbirth, and the puerperium	2.8581	0.2271	0.7521	0.0660	0.2478	1.1065
1998	Diseases of the skin and subcutaneous tissue	1.3847	0.0130	0.0842	0.0305	0.0578	1.1085
1999	Diseases of the skin and subcutaneous tissue	1.5433	0.0137	0.1120	0.0244	0.0673	1.1914
2000	Diseases of the skin and subcutaneous tissue	1.5102	0.0163	0.1172	0.0264	0.0649	1.1728
2001	Diseases of the skin and subcutaneous tissue	1.4430	0.0165	0.1241	0.0373	0.0959	1.1831
2002	Diseases of the skin and subcutaneous tissue	1.5322	0.0196	0.4013	0.0341	0.0993	1.2104
2003	Diseases of the skin and subcutaneous tissue	1.5868	0.0184	0.1295	0.0419	0.0823	1.2065
2004	Diseases of the skin and subcutaneous tissue	1.5200	0.0190	0.1435	0.0417	0.0834	1.1089
1998	Diseases of the musculoskeletal system and connective tissue	3.8286	0.0302	0.2675	0.0444	0.3233	1.1572
1999	Diseases of the musculoskeletal system and connective tissue	3.5155	0.0293	0.2159	0.0380	0.2932	1.2659

continued

TABLE A.3 Continued

Year	Disease	Mean Office Visits	Mean Hospital Admissions	Mean Hospital Number of Nights	Mean Emergency Room Visits	Mean Outpatient Visits	Mean Prescriptions
2000	Diseases of the musculoskeletal system and connective tissue	3.5490	0.0307	0.2023	0.0452	0.2691	1.3400
2001	Diseases of the musculoskeletal system and connective tissue	3.6099	0.0314	0.2001	0.0489	0.2899	1.3520
2002	Diseases of the musculoskeletal system and connective tissue	3.5617	0.0234	0.1281	0.0472	0.3105	1.2897
2003	Diseases of the musculoskeletal system and connective tissue	3.8879	0.0229	0.1229	0.0338	0.2701	1.2703
2004	Diseases of the musculoskeletal system and connective tissue	3.7781	0.0254	0.1324	0.0476	0.2876	1.2979
1998	Congenital anomalies	2.8022	0.1846	0.8865	0.0487	0.1666	0.6433
1999	Congenital anomalies	5.9628	0.1917	1.5292	0.0626	0.2815	0.8122
2000	Congenital anomalies	3.3537	0.1065	0.6976	0.0292	0.7034	0.6986
2001	Congenital anomalies	3.4440	0.1076	0.7964	0.0650	0.9594	0.7943
2002	Congenital anomalies	4.4862	0.1450	1.3944	0.0387	1.4109	0.7631
2003	Congenital anomalies	5.1589	0.1706	0.9526	0.0337	0.4354	0.8738
2004	Congenital anomalies	4.6310	0.1222	0.4082	0.0438	1.1813	0.9655
1998	Injury and poisoning	2.9741	0.0549	0.3398	0.3262	0.2691	0.6217
1999	Injury and poisoning	2.8727	0.0649	0.3563	0.3303	0.3039	0.6548
2000	Injury and poisoning	2.7218	0.0592	0.5439	0.3496	0.2142	0.7107
2001	Injury and poisoning	2.7101	0.0633	0.3878	0.3813	0.3522	0.7172
2002	Injury and poisoning	3.0501	0.0640	0.4327	0.3762	0.3173	0.6753
2003	Injury and poisoning	3.0321	0.0664	0.5234	0.3695	0.3225	0.6875
2004	Injury and poisoning	3.1563	0.0541	0.3164	0.3724	0.2748	0.6568

1998	Other conditions	1.2777	0.0200	0.1612	0.0431	0.1247	1.2794
1999	Other conditions	1.3427	0.0167	0.0853	0.0428	0.0726	1.4015
2000	Other conditions	1.1272	0.0169	0.0990	0.0340	0.1119	1.4490
2001	Other conditions	1.1439	0.0177	0.0710	0.0391	0.0772	1.4752
2002	Other conditions	1.1510	0.0191	0.1086	0.0416	0.1029	1.4686
2003	Other conditions	1.3404	0.0156	0.1066	0.0432	0.1285	1.4266
2004	Other conditions	1.1988	0.0205	0.1015	0.0477	0.1109	1.5059
1998	No diagnosis	1.5179	0.0090	0.0672	0.0054	0.0508	0.1640
1999	No diagnosis	1.4112	0.0067	0.0524	0.0051	0.0339	0.1456
2000	No diagnosis	1.4522	0.0138	0.0463	0.0069	0.0420	0.1369
2001	No diagnosis	1.4922	0.0023	0.0095	0.0114	0.0499	0.1060
2002	No diagnosis	1.5589	0.0044	0.0529	0.0133	0.0607	0.1275
2003	No diagnosis	1.4133	0.0096	0.0452	0.0060	0.0584	0.1157
2004	No diagnosis	1.5486	0.0072	0.0630	0.0109	0.0397	0.1420

SOURCE: Bureau of Labor Statistics handout at the workshop; see text.

TABLE A.4 Total National Medical Expenditures by CPI Stratum and Payee Type (in billions of dollars)

Total Expenditures

	Professional Services		Hospitals		Prescriptions	
Year	Expenditures	Shares	Expenditures	Shares	Expenditures	Shares
1999	165.15	32.15%	256.07	49.85%	92.43	17.99%
2000	177.20	31.94%	273.88	49.36%	103.74	18.70%
2001	213.90	33.18%	295.72	45.87%	135.04	20.95%
2002	248.60	33.62%	338.61	45.79%	152.28	20.59%
2003	254.27	29.20%	437.43	50.23%	179.14	20.57%
2004	293.08	32.48%	416.72	46.19%	192.44	21.33%

Out-of-Pocket Expenditures

	Professional Services		Hospitals		Prescriptions	
Year	Expenditures	Shares	Expenditures	Shares	Expenditures	Shares
1999	22.77	30.80%	9.56	12.93%	41.61	56.27%
2000	22.23	28.19%	9.33	11.83%	47.30	59.98%
2001	28.62	29.42%	9.87	10.15%	58.79	60.44%
2002	33.86	30.92%	11.93	10.90%	63.70	58.18%
2003	32.00	25.84%	12.29	9.92%	79.56	64.24%
2004	36.62	27.96%	13.44	10.26%	80.90	61.77%

All Third-Party Payments

	Professional Services		Hospitals		Prescriptions	
Year	Expenditures	Shares	Expenditures	Shares	Expenditures	Shares
1999	142.38	32.38%	246.50	56.06%	50.82	11.56%
2000	154.97	32.56%	264.55	55.58%	56.44	11.86%
2001	185.28	33.85%	285.85	52.22%	76.25	13.93%
2002	214.74	34.09%	326.68	51.85%	88.58	14.06%
2003	222.27	29.76%	425.15	56.91%	99.57	13.33%
2004	256.46	33.25%	403.28	52.29%	111.53	14.46%

SOURCE: Bureau of Labor Statistics handout at the workshop; see text.

TABLE A.5 Preliminary Results from the Experimental Disease-Based Indexes (CPI January 1999 to December 2004)

Disease	With Updates[a]	Without Updates[a]	With Updates[b]	Without Updates[b]
Infectious and parasitic diseases	66.33	39.28	44.23	29.70
Neoplasms	34.70	42.53	35.34	36.30
Endocrine, nutritional, and metabolic diseases and immunity disorders	46.05	34.68	43.61	29.23
Diseases of the blood and blood-forming organs	20.01	42.63	18.04	30.06
Mental disorders	7.80	37.09	13.08	33.96
Diseases of the nervous system and sense organs	51.21	37.59	37.73	29.34
Diseases of the circulatory system	39.21	41.82	36.77	31.21
Diseases of the respiratory system	37.36	39.40	34.15	32.37
Diseases of the digestive system	24.69	41.82	34.69	33.63
Diseases of the genitourinary system	34.72	40.88	36.81	33.37
Complications of pregnancy, childbirth, and the puerperium	15.92	41.84	21.52	31.64
Diseases of the skin and subcutaneous tissue	71.01	36.03	49.43	29.70
Diseases of the musculoskeletal system and connective tissue	20.89	37.48	29.46	29.37
Congenital anomalies	69.95	43.41	79.66	36.87
Injury and poisoning	63.29	41.71	50.62	36.17
Other conditions	33.02	35.61	39.37	29.58
No diagnosis	21.63	27.14	19.61	27.61

[a]Weighted by total expenditures.
[b]Weighted by out-of-pocket expenditures.
SOURCE: Bureau of Labor Statistics handout at the workshop; see text.

Appendix B

Workshop Agenda and Participants

WORKSHOP AGENDA

Friday, March 14, 2008

8:30 a.m. **Introduction/Overview of Day's Agenda**
Joseph Newhouse, Steering Committee Chair

8:45 **Plans for a Satellite Health Care Account.** Goals of BEA's health care accounting program; progress to date on producing a detailed proposal by the end of 2009; summary of BEA's strategies for dealing with key measurement issues; and data needs.
Ana Aizcorbe, BEA
Discussants: *Dale Jorgenson, Matthew Shapiro*
Open Discussion

10:00 **Constructing Nominal Expenditures by Disease.** Overview of Altarum Institute's work developing time-series estimates of national expenditures by medical condition. These estimates are benchmarked to government estimates of personal health expenditures from the National Health Expenditure Accounts. Thus, they cover spending by the civilian noninstitutional population as well as some institutionalized populations and the military.
Charles Roehrig, Altarum Institute
Discussants: *David Cutler, Allison Rosen*
Open Discussion

APPENDIX B
85

11:15   **Coffee Break**

11:30   **Price Indexes and Volume Measures.** Discuss BLS's plans to research and generate price indexes organized by broad disease category. The PPI program currently publishes provider based price indexes aggregated by broad disease category for payers other than Medicare and Medicaid in general hospitals and for pharmaceutical manufacturing. Because similar data will be available in the future for other industries such as physician services, medical laboratories, and diagnostic imaging centers, the PPI is researching the development of an alternative price index organized by disease categories rather than provider. Currently collected item data captured monthly and priced at the point of service (the provider) will be used in the proposed index, and procedures have been developed that allow price change caused by substitution of treatments across providers to be reflected. Additionally, PPI has developed a method to quality adjust its current hospital indexes by using quality indicators contained in the CMS Hospital Compare database. The CPI program is generating experimental price indexes, also organized by major disease category, by merging medical expenditure and utilization data from the Medical Expenditure Panel Survey with the BLS CPI production database. Both the PPI quality adjustment proposal and the CPI generation of experimental price indexes use existing databases with no additional expenditure from BLS.

    Bonnie Murphy, BLS
    Ralph Bradley, BLS
    Discussants: *Jack Triplett, Patricia Danzon*
    Open Discussion

12:45 p.m.   **Lunch Topic Discussion: Measuring Treatment Outcomes.** Discuss the difficulties when constructing measures of treatment outcomes and assess the current state of knowledge. Discuss the role of clinical data to inform performance measures; offer views on merging clinical data to claims data, and on the availability of HMO encounter data.

    Mark McClellan, Brookings Institution
    Open Discussion

2:00   **National Accounting Issues.** Discuss national accounting issues that must be resolved in order to produce a satellite account for health care. Among these is how to construct measures of real expenditures for health care industries, define disease and product classes, and make the spending and industry sides of the account

add up (or provide a reason why they do not). Also discuss the need for a set of deflators for industries (at least doctors, hospitals and drugs); the importance of data on sources of payment, and the role of government subsidies; the domain for goods and services, including a time consideration—at what point do we draw the line for treatment; what to include in gross output or final demand; and how to deal with nonmarket health inputs.
Brian Moyer, BEA
Discussants: *Barbara Fraumeni, Sherry Glied*
Open Discussion

3:15 **Summary and Perspective, Prospects for Moving Forward.** Wrap-up comments on (1) BEA's plans for the NIPAs and for the satellite health care accounts, (2) hurdles and priorities, and (3) next steps.
Gail Wilensky, Project HOPE
Open Discussion

4:00 p.m. **Adjourn**

LIST OF PARTICIPANTS

Ana Aizcorbe, BEA
Bruce Baker, BEA
Jessica Banthin, AHRQ
Linda Bilheimer, NCHS
Ralph Bradley, BLS
LeRoynda Brooks, BEA
Elaine Cardenas, BLS
Aaron Catlin, CMS
Constance Citro, CNSTAT Staff
Steve Cohen, AHRQ
Frank Congelio, BLS
Cathy Cowan, CMS
David Cutler, Harvard University
Patricia Danzon, University of Pennsylvania
Dennis Fixler, BEA
Barbara Fraumeni, University of Southern Maine
Mark Freeland, CMS
Alan Garber, Stanford University
Daniel Ginsburg, BLS
Sherry Glied, Columbia University
Micah Hartman, CMS

Steven Heffler, CMS
Michael Horrigan, BLS
Dale Jorgenson, Harvard University
Richard Kane, BEA
Emmett Keeler, RAND
James Kim, BEA
Steven Landefeld, BEA
Katharine Levit, Thomson Healthcare
John Lucier, BLS
Erin Ludlow, BEA
Christopher Mackie, CNSTAT Staff
William Marder, Thomson Healthcare
Tami Mark, Thomson Healthcare
Mark McClellan, Brookings Institution
Robert McClelland, BLS
Brent Moulton, BEA
Brian Moyer, BEA
Bonnie Murphy, BLS
Joseph Newhouse, Harvard University
Sarah Pack, BEA
Bonnie Retus, BEA
Charles Roehrig, Altarum Institute
Allison Rosen, University of Michigan
Arthur Sensenig, CMS
Matthew Shapiro, University of Michigan
Michael Siri, CNSTAT Staff
Shelly Smith, BEA
Edward Sondik, NCHS
Jack Triplett, Brookings Institution
Frank Velez, BLS
Gail Wilensky, Project HOPE

Appendix C

Adapting BEA's National and Industry Accounts for a Health Care Satellite Account

Brent R. Moulton, Brian C. Moyer, and Ana Aizcorbe

INTRODUCTION

The Bureau of Economic Analysis (BEA) covers the production of health care goods and services in its national income and product accounts (NIPAs) and in its industry accounts. The NIPAs include estimates of nominal and real spending by consumers and government on health care goods and services, and the industry accounts include nominal and real estimates of output, intermediate inputs, and value added for the health care industries. As BEA begins to think about a possible health care satellite account, some important modifications to the existing framework underlying the NIPAs and the industry accounts may be required to emphasize the interrelated nature of health care provision and to facilitate the use of improved price indexes.

An important aspect of developing a health care satellite account involves a change in the definition of the final good(s) provided by the health sector from the individual treatments to the provision of "medical care." Using the latter definition, the BEA satellite account will use disease-based price indexes to deflate consumer spending on medical care and thus potentially change the growth rate of real gross domestic product (GDP).

This paper discusses how BEA's accounts might be modified to accommodate this new definition. The delivery of medical care generally requires the coordinated provision of goods and services by several providers. BEA's accounts have traditionally focused on separately measuring the output of each type of provider (e.g., physicians, hospitals, outpatient facilities, pharmaceutical manufacturers and distributors, etc.). Consequently, the accounts do not directly measure the improvements that are possible through substituting or more efficiently combining the various modes of service. We suggest a modified framework in which a

APPENDIX C   89

physician orchestrates and manages patients' medical care by making diagnoses and pointing the patient to other providers for procedures, lab work, and the like. The services provided by these other providers would be viewed as intermediate goods and services in the provision of the final output, medical care. The advantage of adopting this view of the health sector is that it provides a natural way to accommodate the new definition of the "good" through standard double-deflation methods. An important side benefit is that the new structure provides a role for both disease-based price indexes—to deflate nominal spending—and the Bureau of Labor Statistics' Producer Price Indexes (PPIs)—to deflate the intermediate goods.[1]

REROUTING OF HEALTH CARE TRANSACTIONS IN BEA'S ACCOUNTS

The financing of health care, whether by private health insurance or government social insurance funds, involves complicated transactions. The standard presentation of BEA's core accounts already involves *rerouting* transactions—that is, recording transactions as taking place through channels different from the ones through which they actually occur—to identify the economic purpose of these transactions.[2] As part of developing a health care satellite account, some different forms of rerouting are likely to be required.

Some of the actual transactions for health care provided through an employer-provided traditional health insurance plan are shown in Figure C.1. Typically, both the employer and the employee pay premiums into the plan. The employee and his or her family then obtain goods or services from various health care providers. The health plan pays an agreed-upon portion of the cost to the provider, and the employee also pays copayments and deductibles.

BEA's accounts, in contrast, show the entire employer contribution as part of labor cost (compensation of employees) for the firm and as part of personal income for the employee (Figure C.2). All of the purchases of health care are shown as purchases by households, representing the ultimate consumer of the health care goods and services, rather than as shared purchases by the household and the health insurance plan. In the economic accounts, the principal role of the health insurance plan is as a provider of health insurance services, an imputed transaction equal in value to the difference between premiums and expected benefits, which is treated as a service purchased by the covered employees.

A similar rerouting of transactions is associated with Medicare Part A (hospital insurance). (Other types of health care funding, such as Medicare Parts B,

---

[1] Aizcorbe and Nestoriak (2007) used this framework to interpret differences in disease-based and treatment-based price indexes.

[2] See Commission of the European Communities, International Monetary Fund, Organisation for Economic Co-operation and Development, United Nations, and the World Bank (1993, paragraphs 3.24-3.27).

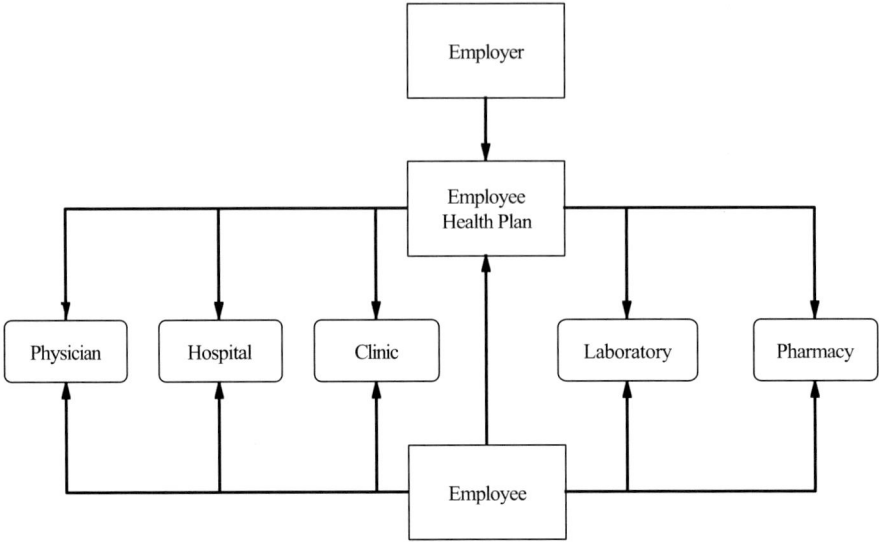

FIGURE C.1 Funding of health care, employer-provided health insurance.
SOURCE: Paper prepared for the Health Accounting Workshop by Brent R. Moulton, Brian C. Moyer, and Ana Aizcorbe.

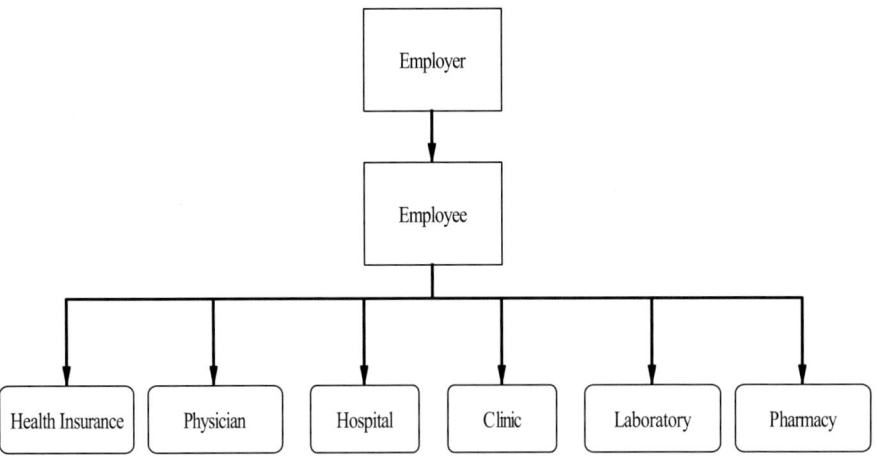

FIGURE C.2 Private employer-provided health insurance after rerouting.
SOURCE: Paper prepared for the Health Accounting Workshop by Brent R. Moulton, Brian C. Moyer, and Ana Aizcorbe.

APPENDIX C 91

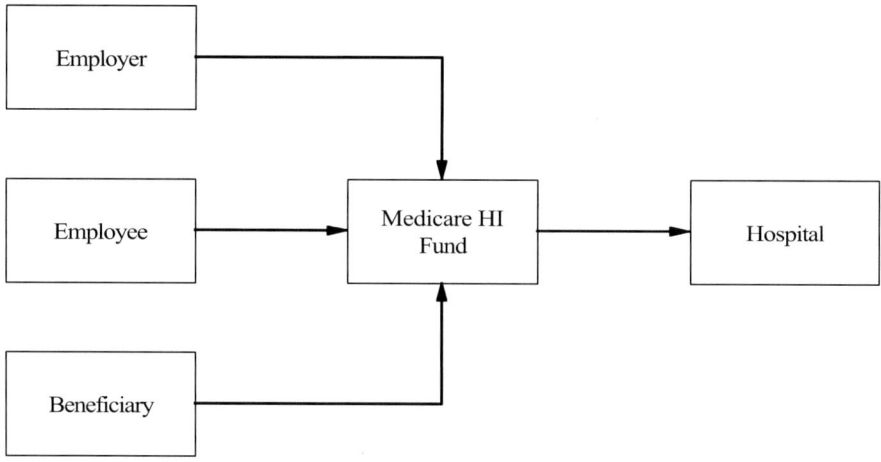

FIGURE C.3 Funding of Medicare Part A (hospitalization insurance).
SOURCE: Paper prepared for the Health Accounting Workshop by Brent R. Moulton, Brian C. Moyer, and Ana Aizcorbe.

C, and D and Medicaid, exhibit similar rerouting of transactions.) As shown in Figure C.3, the Medicare hospital insurance fund is financed largely through employer and employee contributions (payroll taxes). Note that in contrast to the last example, in most cases the employee is not a current beneficiary of the program and therefore is not the consumer of the health care, although the employee and employer contributions provide for future eligibility. Most beneficiaries are not required to pay premiums for Part A, but some individuals who are not otherwise eligible pay premiums to buy coverage; thus they are another source of funding for the Medicare HI (hospitalization insurance) Fund. In most cases, Medicare Part A pays for covered medical services (primarily hospital inpatient services and inpatient services in skilled nursing facilities) without requiring copayments or deductibles.

After rerouting, the Medicare Part A transactions take the form shown in Figure C.4.[3] The employer contributions are counted as part of the compensation of employees, so that they are included in the enterprise's labor costs. They are then shown as contributed to the Medicare fund, so that they are not included in personal income. The value of the benefits is shown as a transfer (social benefits) to persons and included in personal income. The consumption of health services by covered individuals is recorded as an imputed purchase of health services in

---

[3] The treatment of government-funded health care is discussed in Bureau of Economic Analysis (2005).

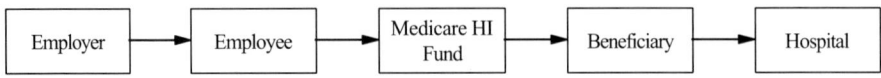

FIGURE C.4 Medicare Part A after rerouting.
SOURCE: Paper prepared for the Health Accounting Workshop by Brent R. Moulton, Brian C. Moyer, and Ana Aizcorbe.

personal consumption expenditures. The administrative costs of the Medicare program are included in government consumption expenditures.

SATELLITE HEALTH CARE ACCOUNT

One of the key features of a BEA health care satellite account will be the classification of health expenditures by type of disease or condition rather than by type of provider. This will allow for the use of deflators organized by type of disease and therefore that better capture substitution across types of providers—for example, substituting outpatient for inpatient hospital treatment or substituting pharmaceuticals for more invasive treatments, such as surgery. The disease-based approach will also allow BEA to focus on the costs and benefits of treatments for specific types of diseases.

To have a comprehensive accounting of the productivity gains from this type of substitution, the gains must be attributed to one or more of the provider industries. One simple possibility would be to simply allocate the productivity gains across industries, assuming that they all contribute proportionally to the gains. However, we note that physicians may play an especially important role, since they tend to serve as managers and decision makers in combining the goods and services of various providers in producing medical care. For example, physicians tend to make decisions about what lab tests to run, when hospital services are needed, and so forth. That suggests another approach that BEA is currently investigating, the possibility of rerouting existing health care transactions through the physician services industry, whose output can then be classified by products defined along lines of type of disease.

Consider an example in which the management services are provided by a primary caregiver. (Depending on the type of care, the manager/decision maker may be a physician specialist or a nonphysician medical professional.) Comparing with Figure C.2, there is a rerouting of transactions to create a primary caregiver who then treats each of the other types of providers as an intermediate input to the caregiver's production. The notion underlying this modification to the existing framework is that patients have a primary caregiver who acts as a manager in orchestrating patients' medical care. This is the type of organization used, for example, by health maintenance organizations, which consolidate all types of services so that customers transact with a single organization with respect to copayments or other billing. In many cases, it seems reasonable to think of other providers as performing an intermediate role to the primary caregiver. For example, for lab work associated with a routine office visit, the patient probably has no direct interactions with the lab and probably does not know the identity of the lab until the bill arrives; it seems a bit anachronistic that the billing is done separately, rather than being charged through the physician who ordered the lab work. For other types of providers, the patient may exercise more discretion—for example, the patient may choose a pharmacy based on price or convenience, but the physician controls what drug is prescribed. Similarly, a physician may or may not offer a patient a choice of hospitals when an inpatient stay is required. These examples suggest that the relationship between the primary caregiver and other providers may have important similarities to the typical general relationship between a producer and the providers of intermediate inputs. Figure C.5 illustrates the rerouting that may be used in this case.

This proposed modification to BEA's accounting framework would have no direct impact on the aggregate estimates of consumer spending in the NIPAs (the detailed estimates would be presented by type of disease). It would, however, impact the estimates of real consumer spending, and therefore, real GDP. Under the modified framework, consumer services provided by the physician

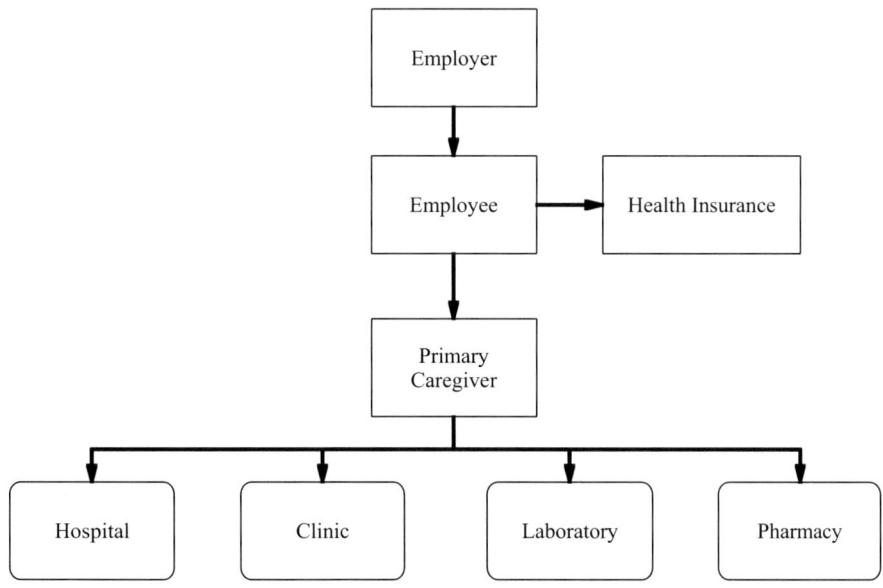

FIGURE C.5 Consolidated health sector.
SOURCE: Paper prepared for the Health Accounting Workshop by Brent R. Moulton, Brian C. Moyer, and Ana Aizcorbe.

or other professional serving as manager (defined by type of disease) would be deflated using disease-based price indexes, which better capture substitution across types of providers. This is in contrast to BEA's current framework (presented in Figure C.2), which relies on PPIs to deflate the goods and services for specific health care providers.

Within BEA's industry accounts, the modified framework would introduce a new, primary caregiver industry that would subsume the existing industry, "offices of physicians." The output of this new industry would include the value of the intermediate inputs purchased from the individual health-care-providing industries and the value added of offices of physicians. The output of the consolidated health care industry would then be deflated using disease-based price indexes, while its intermediate inputs would be deflated using PPIs. Real value added—computed using the double-deflation method as the difference between real output and real intermediate inputs—would reflect this new industry's contribution to real GDP, including industry productivity gains. One can think of a health care system that facilitates the diffusion of new goods by providing information on new treatments. When these efforts successfully prompt the primary caregiver to prescribe different, lower cost treatments, this is reflected in the real value added of the consolidated health care industry.

ONGOING AND FUTURE WORK

BEA is in the beginning stages of developing a health care satellite account. As discussed in this paper, efforts are under way to identify how existing accounting frameworks can be adapted to best suit a satellite account. Efforts are also under way to develop disease-based estimates of health care spending using private insurance claims data, Centers for Medicare and Medicaid Services data on Medicare and Medicaid recipients, and data on the uninsured from the U.S. Department of Health and Human Services. In addition, BEA is developing disease-based price indexes that will be used to deflate these new nominal health expenditures.

When complete, BEA's health care satellite account will generate measures of health care spending that can be used to better track the sources of rising health care costs. In addition, BEA is working with economists and health care experts to explore ways that these cost measures may be integrated with models of disease prevalence and health status in order to better assess the potential benefits of spending on health care.

REFERENCES

Aizcorbe, A., and Nicole Nestoriak. (2007). *Changes in Treatment Intensity, Treatment Substitution, and Price Indexes for Health Care Services.* Paper presented at the National Bureau of Economic Research Productivity Workshop, December 5, Cambridge, MA.

Bureau of Economic Analysis. (2005). *MP-5, Government Transactions.* (September). Washington, DC: U.S. Department of Commerce.

Commission of the European Communities, International Monetary Fund, Organisation for Economic Co-operation and Development, United Nations, and World Bank. (1993). *System of National Accounts 1993.* New York: United Nations.

COMMITTEE ON NATIONAL STATISTICS

The Committee on National Statistics (CNSTAT) was established in 1972 at the National Academies to improve the statistical methods and information on which public policy decisions are based. The committee carries out studies, workshops, and other activities to foster better measures and fuller understanding of the economy, the environment, public health, crime education, immigration, poverty, welfare, and other public policy issues. It also evaluates ongoing statistical programs and tracks the statistical policy and coordinating activities of the federal government, serving a unique role at the intersection of statistics and public policy. The committee's work is supported by a consortium of federal agencies through a National Science Foundation grant.